NUNC COGNOSCO EX PARTE

THOMAS J. BATA LIBRARY
TRENT UNIVERSITY

PROBABILITY AND THE
WEIGHING OF EVIDENCE

Other books of interest
published by Charles Griffin & Co. Ltd.

★

By G. U. YULE *and* M. G. KENDALL
 Introduction to the theory of statistics

By M. G. KENDALL
 The advanced theory of statistics (2 vols.)
 Rank correlation methods

By F. YATES
 Sampling methods for censuses and surveys

PROBABILITY AND THE WEIGHING OF EVIDENCE

By

I. J. GOOD, M.A., Ph.D.

FORMER LECTURER IN MATHEMATICS
AT THE UNIVERSITY OF MANCHESTER

LONDON
CHARLES GRIFFIN & COMPANY LIMITED

NEW YORK
HAFNER PUBLISHING COMPANY

PREFACE

"Probability is the very guide of life."
CICERO, *De Natura*

WHEN we wish to decide whether to adopt a particular course of action, our decision clearly depends on the values to us of the possible alternative consequences. A rational decision depends also on our degrees of belief that each of the alternatives will occur. Probability, as it is understood here, is the logic (rather than the psychology) of degrees of belief and of their possible modification in the light of experience.

The aim of the present work is to provide a consistent theory of probability that is mathematically simple, logically sound and adequate as a basis for scientific induction, for statistics, and for ordinary reasoning. Probability is treated as a subject in its own right, of comparable importance to the related subjects of philosophy, statistics and mathematics. I hope there is not a disproportionate stress on either the philosophical, statistical or mathematical aspects.

Various authorities have attempted to eliminate the necessity for subjective probability judgments by employing instructions that are outside the theory adopted here. These instructions are either imprecisely stated or, when they are precise, apply only to ideal circumstances, so that they can be used only in some unspecified approximate sense. The instructions occasionally contradict one's inner convictions. It is maintained here that judgments should be given a recognised place from the start. These judgments are influenced by a free discussion of standard instructions, but they are not bound by them.

The necessity for judgments occurs most conspicuously in connexion with "initial probabilities" of hypotheses. When a scientific memoir is concerned with experimental evidence for a hypothesis, it is helpful if something is stated about the subjective initial probability of the hypothesis. To omit such a statement gives only a superficial appearance of objectivity. The uninitiated are liable to be misled into regarding the probability as higher than would be claimed by the writer of the memoir.

The theory presented in the following pages follows precise rules, although it uses subjective judgments as its raw material. In this respect it resembles any other scientific theory. But the analogy with other scientific theories should not be pressed too far, since probability is a part of reasoning and is therefore more fundamental than most theories.

Although probability cannot be defined entirely within the framework of formal logic and pure mathematics it is possible to go some way in this direction by adopting the axiomatic method. This method makes it possible to prove many mathematical theorems that are connected with probability, but it does

PREFACE

not explain how these theorems are to be used. For this purpose some philosophical interpretation of probability is required.

A condensed account is given in Chapter 1 of various theories of probability which have been suggested in the past, together with some brief criticisms. In Chapters 2 and 3 the axiomatic part of the present theory is developed. Chapter 4 is more philosophical. It deals with the rules of application of the abstract theory developed in Chapters 2 and 3. Some of the questions are difficult and the answers are not entirely satisfactory, but other theories do not seem to have given better answers. In this chapter the apparent dualism of probability is attributed to the use of different kinds of propositions rather than to different kinds of probability. This point of view is largely responsible for the extreme simplicity of the formal apparatus, in spite of the generality of the theory. Chapter 5 provides a background of elementary statistics and probability, sufficient for later use. A few important theorems are quoted without proof. In Chapter 6 the intuitive idea of weighing evidence is given a simple quantitative interpretation. For this purpose it is found convenient to use the term " plausibility " for the logarithm of odds. A gain of plausibility bears about the same relation to an " amount of evidence " as a probability bears to a " degree of belief ". The term is used in the discussion of statistics in the last chapter.

The following is a list of ordinary words that are generally used in this book in a technical sense (roughly in order of their appearance) : *belief, you* (this is always a technical term), *comparison, theory of probability, body of beliefs, reasoning, reasonable, contradiction* (in a body of beliefs), *probability, abstract theory, rules, impossible, certain, almost, independence, theory* (meaning " hypothesis " or " scientific theory " or an abbreviation for " theory of probability "), *proper theory, improper theory.*

The use of the word " reasonable " as a technical term is intended to be partly emotive—it involves a recommendation to use the theory in practice. Otherwise the theory would be tautological in the sense in which pure mathematics is tautological.

I have of course been much influenced, directly and indirectly, by many other writers, and especially by F. P. Ramsey, H. Jeffreys, B. O. Koopman, R. von Mises, J. M. Keynes and A. Kolmogoroff. I am indebted also to Dr. A. M. Turing and Professor M. S. Bartlett with whom I have had several illuminating conversations. After reading the manuscript, Professor Bartlett felt that the treatment was not always quite fair to the orthodox statistical theory and I have attempted to rectify this. Dr. A. M. Turing, Professor M. H. A. Newman and Mr. D. Michie were good enough to read the first draft (written in 1946) and I am most grateful for their numerous suggestions. I am grateful also to the publishers who have been most helpful at every stage.

December, 1949
I. J. GOOD

LIST OF CHAPTERS AND SECTIONS

Preface v

THEORIES OF PROBABILITY

1.1	Logical notation	1
1.2	Degrees of belief	1
1.3	Purposes of a theory of probability	3
1.3A	The "axiomatic" method	5
1.4	Some theories of probability	6

THE ORIGIN OF THE AXIOMS

2.1	Preamble	13
2.2	Two "obvious" axioms	13
2.3	Definition of numerical probability by judgment of equally probable alternatives	14
2.4	Example	15
2.5	The law of addition of probabilities	16
2.6	The law of multiplication of probabilities	16
2.7	Example	17
2.8	Continuous probabilities	17

THE ABSTRACT THEORY

3.1	The axioms	19
3.2	Definitions	21
3.3	Theorems	22
3.4	An alternative set of axioms	30

THE THEORY AND TECHNIQUE OF PROBABILITY

4.1	The "rules"	31
4.1A	The justification of the theory	33
4.2	Inaccurate language	33
4.3	Some "suggestions"	34
4.4	A non-numerical theory	36
4.5	Practical difficulties	36
4.6	The principles of "insufficient reason" and "cogent reason" ..	36
4.7	Simple examples	38
4.8	Certainty and the "verification" of the theory	39
4.9	Deciding between alternative hypotheses or scientific theories ..	40
4.10	Connexions with the frequency theory	46
4.11	Relation to the objective theory	47
4.12	Generalisation of $\mathcal{B}$	48
4.13	Degrees of belief concerning mathematical theorems	49
4.14	Development of the judgment by betting	49

LIST OF CHAPTERS AND SECTIONS

page

PROBABILITY DISTRIBUTIONS

5.1	Random variables and probability distributions	50
5.2	Expectation	52
5.3	Examples of distributions	55
5.4	Statistical populations and frequency distributions	59

WEIGHING EVIDENCE

6.1	Factors and likelihoods	62
6.2	"Sequential tests" of statistical hypotheses	64
6.3	Three hypotheses and legal applications	66
6.4	Small probabilities in everyday life	68
6.5	Composite hypotheses	68
6.6	Relative factors and relative probabilities	71
6.7	Expected weight of evidence	72
6.8	Exercises	73
6.9	Entropy	74

STATISTICS AND PROBABILITY

7.1	Introduction	76
7.2	Sampling of a single attribute	77
7.3	Example (ESP again)	81
7.4	Inverse probability versus "precision"	82
7.5	Sampling and the probabilities of chance distributions (curve-fitting)	84
7.6	Further remarks on curve-fitting	88
7.7	Combination of observations	89
7.8	Significance tests	90
7.8A	The chi-squared test	93
7.8B	Additional note on the chi-squared test	95
7.9	Contingency tables	97
7.10	Estimation problems	101

Appendices

I	The error function	105
II	Dirichlet's multiple integral	105
III	On the conventionality of the addition and product laws	105

References .. 107

Index .. 109

CHAPTER 1

THEORIES OF PROBABILITY

"I would rather feel compunction than understand the definition thereof." THOMAS À KEMPIS

1.1 Logical notation

We shall not delve deeply into ordinary logic. The symbols E, F, G, H, E' etc. will denote propositions. A proposition is defined † to be a statement for which it is meaningful to assert that it is true or that it is false. (The meaning of "meaning" will not be discussed.‡) A proposition may be simple or complicated, it may refer to past, present or future and to a real or imaginary world. It will not contain a reference to probability, at any rate not before probability has been defined.

The negation of E will be denoted by "$\bar{E}$" (read "not E"); the conjunction of E and F by "$E.F$" (read "E and F"), and the disjunction by "$E \vee F$" (read "E or F"). The disjunction is true if either E or F or both are true. The notation may be extended to conjunctions and disjunctions of more than two propositions.

1.2 Degrees of belief

Our theory of probability is concerned with those mental phenomena called "degrees of belief" (i.e. "states" of more or less belief). Some people use the word "belief" in a sense which precludes the use of the phrase "degree of belief". They would say that they either believe so-and-so or that they do not. That it is sensible, however, to talk about degrees of belief, at any rate in some circumstances, can be shown by considering a simple example. My belief that it will rain to-morrow is more intense than my belief that the roof above me will collapse. To say that the first degree of belief is greater than the second is another way of saying the same thing. To prevent misunderstanding it may be noticed further that to say that one degree of belief is more intense than another is not intended to mean that there is more emotion attached to it. What is meant is sufficiently shown by the above example: a complete definition can hardly be produced.

It will not be assumed at the outset that degrees of belief can be measured

† See Hilbert and Ackermann, 1946, 3. (The references are at the end of the book.)
‡ It seems to the writer that there are "degrees of meaning" and hence that there are sentences for which it is difficult to decide whether they are propositions. Such "partial propositions" often occur in the pioneering work on new scientific theories.

numerically, in spite of the word "degree". For short they will often be called simply "beliefs". A belief depends very roughly on three variables: the proposition "believed" (say E), the proposition assumed (say H),† and the general state of mind ($\mathfrak{M}$) of the person who is doing the believing. This person will be described as "*you*". $\mathfrak{M}$ depends on who "you" are and on the moment of believing. It will be convenient to use the symbol $B(E \mid H : \mathfrak{M})$ for this belief, and it may be read "your (degree of) belief in E if H is assumed, when your state of mind is $\mathfrak{M}$". It will be written $B(E \mid H)$ when $\mathfrak{M}$ is taken for granted. It is important to realise that H need not be known to be true; $B(E \mid H : \mathfrak{M})$ is your estimate, when your state of mind is $\mathfrak{M}$, of what your degree of belief in E would be if you knew H to be true.

As an example suppose that H is deducible from ordinary logic (i.e. it is an "analytic proposition") and that E is an empirical proposition about the material world. It is then by no means obvious that any meaning can be attached to $B(E \mid H : \mathfrak{M})$. In order to feel convinced that $B(E \mid H : \mathfrak{M})$ has a meaning when E is empirical most people would consider that H also should involve a certain amount of empirical information. It will be assumed at any rate that $B(E \mid H : \mathfrak{M})$ does sometimes mean something.

A belief $B(E \mid H : \mathfrak{M})$ is subjective in the sense that it depends on $\mathfrak{M}$. Keynes and Jeffreys ‡ assume that there is a "reasonable" (degree of) belief which is independent of $\mathfrak{M}$. This may be called an "objective" belief.§ They call it a probability and it depends only on E and H. The notation used by Jeffreys is $P(E \mid H)$. This meaning for "probability" is not quite the same as the one that will be adopted here. It is true that a probability will soon be defined roughly as a reasonable belief, but it will be maintained that reasonableness does not necessarily imply complete objectivity.

It is perhaps hardly necessary to admit that no precise definition will be given of a belief. Instead it will be taken as a primitive notion. The present work may be regarded as an analysis of properties of this notion rather than as a definition.

It is possible for one of your beliefs $B(E \mid H : \mathfrak{M})$ at a given time to be *more intense* than another one $B(E' \mid H' : \mathfrak{M}')$ at some other time. This too will be taken as a primitive notion and will be denoted by $B(E \mid H : \mathfrak{M}) > B(E' \mid H' : \mathfrak{M}')$ or by $B(E' \mid H' : \mathfrak{M}') < B(E \mid H : \mathfrak{M})$. The symbols "$>$" and "$<$" may

---

† E often asserts that an event has happened or will happen, while H is often regarded as a hypothesis. But this is unnecessary: we regard E and H as arbitrary propositions.

‡ See Keynes, 1921, and, for example, Jeffreys, 1939.

§ The words "subjective" and "objective", when applied to theories of probability, have often been used to mean theories that depend respectively on degrees of belief and on the idea of frequency. These words will not be used here in this way.

An objective degree of reasonable belief is called a "credibility" by Bertrand Russell in *Human Knowledge* (London, 1948).

be read " is more intense than " and " is less intense than ", respectively. It will not be assumed that any two beliefs can be compared in this way, even though they are both associated with the same person. Similarly if there are examples of equal intensity the symbol " $=$ " will be used. An " inequality " or " equality " between beliefs will be called a *comparison* between beliefs. Such a comparison, unlike a single belief, is expressed by a sentence containing a verb. There may be no objection to regarding it as a proposition, but the point is not of immediate importance.

1.3 Purposes of a theory of probability

Ordinary logic seems to be inadequate by itself to cope with problems involving beliefs. In addition a theory of probability is required. Such a theory is defined here as a fixed method which, when combined with ordinary logic, enables one to draw deductions from a set of comparisons between beliefs and thereby to form new comparisons.† A set of comparisons between beliefs will be called a *body of beliefs* and will be denoted by a symbol such as " $\mathcal{B}$ " or " $\mathcal{B}'$ ". Thus the immediate purpose ‡ of a theory of probability is to enlarge $\mathcal{B}$.

A fixed theory of probability together with a fixed theory of logic will be called *reasoning*.

A *reasonable* $\mathcal{B}$ will be defined as one such that when it is submitted to the processes of reasoning no *contradiction* emerges. By a " contradiction " is meant here a pair of comparisons that are formally contradictory when the $\mathfrak{M}$'s are omitted, e.g.

$$B(E \mid H : \mathfrak{M}_1) > B(E' \mid H' : \mathfrak{M}_2), \qquad B(E \mid H : \mathfrak{M}_3) < B(E' \mid H' : \mathfrak{M}_4).$$

Observe that the meaning of " reasonable " depends on the system of reasoning and in particular on the theory of probability that is used. The use of the word may therefore be regarded as consistent with ordinary usage if and only if the system of reasoning is itself reasonable in an ordinary sense. A necessary condition for this is that the longest period of time between any pair of the $\mathfrak{M}$'s must not be too great. It is hardly to be expected that " your " judgments would remain quite constant over a long period of time. But if the periods which are involved are short, then the sort of consistency mentioned is a natural requirement.

The beliefs involved in a reasonable $\mathcal{B}$ will be called *probabilities* § and

† Cf. Koopman, 1940. The phrase " a theory of probability " will also be used with its ordinary vague meaning, and which meaning is intended should be clear from the context.

‡ The question of how probability may be used as a guide to rational behaviour will be considered in 5.2.

§ If there are any meaningless symbols $B(E \mid H)$ the corresponding probabilities may be given conventional meanings. Thus a probability is *a reasonable belief if there is one, and is otherwise something introduced for theoretical convenience.*

the symbol B will be replaced by P. The symbol $\mathfrak{M}$ will be omitted so that we are back to Jeffreys' notation $P(E \mid H)$. The use of this notation does not imply that a probability is independent of who "you" are. In any given application "you" are supposed to remain the same person throughout. When it is desired to bring $\mathfrak{B}$ into evidence $P_\mathfrak{B}(E \mid H)$ may be written instead of $P(E \mid H)$. The particular theory of probability will not be mentioned in the notation.

We shall assume that a $\mathfrak{B}$ is reasonable until it proves to be unreasonable. So we shall always use the symbol P rather than B, though this notation is strictly justified only for beliefs involved in a reasonable $\mathfrak{B}$. If a contradiction is reached it may mean that $\mathfrak{B}$ has been too hastily formulated and that it contains a comparison that can be crossed out on more mature consideration.

The comparisons in a body of beliefs are bound to be subjective judgments if no theory of probability has been applied. They may be called *probability judgments* (if it is assumed that $\mathfrak{B}$ is reasonable). The possibility of probability judgments of a more general type will be discussed in Section **4.12**.

A probability theory, being a fixed procedure, lends a *certain amount of objectivity to your subjective beliefs*. If comparisons can be deduced from a $\mathfrak{B}$ that is "empty" (i.e. contains no comparisons) then the comparisons may be described as *objective*.† Similarly an objective theory of probability is one that is designed to work with empty bodies of belief, i.e. without using bodies of belief at all. It seems unlikely to the present writer that a generally applicable objective theory can be constructed,‡ in spite of claims which others have implicitly made. (It should perhaps be emphasised that the phrase "theory of probability" is here being used in the sense defined at the beginning of this section.)

An analogy can be drawn with formal logic, in which new propositions can be deduced from a given body of propositions. In geometry, new relations between points, lines and planes can be deduced from a given set of such relations. A similar property is possessed by all scientific theories.

In order to build up your beliefs it is theoretically sufficient to use reasoning only, without collecting empirical information.§ But in practice this would take too much time: you may be interested in whether E is true but not interested in $P(E \mid H)$ until H becomes an observational fact.

---

† Perhaps a better description would be "constructibly" objective.

‡ It would first be necessary to invent a special language in which statements could be made without any ambiguity of meaning. In ordinary language such statements are rare and perhaps non-existent. (See also **4.11**.)

§ But some experience of the real world may be required in order to understand the meanings of E and H.

1.3A The "axiomatic" method

It is advisable to digress for a moment in order to discuss what is meant by the "axiomatic" method in mathematics. It consists in stating a number of assumed relations between various things which are denoted by words or symbols. These relations are called "axioms", and all the mathematical results are deduced from them. In the course of these deductions no use is made of the *meanings* of the words or symbols; in fact, it is unnecessary to assume that they have any meanings. The position is different when the theory is applied to practical problems.

The method has been successful in all branches of mathematics and in formal logic. Its advantages are that the mathematics depends only on mathematical assumptions and that new assumptions, either mathematical or non-mathematical, are prevented from creeping in. The axioms are often born in some concrete interpretation of the undefined words or symbols. But the structure is strengthened by cutting it away from its origins, since the number of assumptions is thereby decreased.

The method will be adopted here for the treatment of probability. The development from the axioms alone will be called *the abstract theory*. Besides the axioms it is necessary to have a set of rules by which the abstract theory may be applied. The word "rules" will nearly always be used in this sense. An axiomatic theory should always be supplemented by a set of clearly stated rules, if it is to be directly applicable. This condition has not often been satisfied in the past.

The question arises how to select a suitable theory. It must be logically consistent and, more generally, it must never force you into a position that after mature consideration you regard as untenable. (This would happen if a body of beliefs became classified as "unreasonable" while not containing any judgments that could be conscientiously removed.) The theory should be applicable to most of the practical problems concerning degrees of belief, and it would be convenient for it to apply also to idealised problems.†

If the axiomatic method is used it is advisable that the axioms should be simple and should involve a minimum of assumptions. In order to arrive at a system of axioms the classical theories may be used as a guide, especially as it is known that these theories have led to much the same general structure for the subject as a whole, though not always by strictly logical steps. Hence it will be convenient at this point to consider some well-known theories.

---

† This last condition will be partially sacrificed in order that the axioms should involve fewer assumptions. (See the remarks about "complete additivity" in Section **3.3**, pages 22–3.)

1.4 Some theories of probability

Theories of probability may be cross-classified in at least four ways:—

(a) The theory may or may not be dependent on a system of axioms.

(b) Each probability may or may not be defined, or assumed to exist, objectively, i.e. independently of the views of particular people.

(c) The emphasis may be on degrees of belief or on the frequency with which things happen. In the latter case the theory is normally described as a frequency or statistical theory.

(d) Probabilities may or may not be associated with numbers.

Several special theories will now be considered. There are many others, but the ones outlined are fairly representative. My intention is to give a good general picture rather than to mention all the important work. The classifications following each heading are supposed to be those which the adherents of the theories would accept.

(i) *The Venn limit.*† (Classification: non-axiomatic, objective, statistical, numerical.) Imagine that an experiment ‡ or "trial" is repeated an infinite number of times. Then the probability of a "success" is defined as the limit of the proportion of successes in the first n trials when $n \to \infty$. It is assumed that the limit exists. Of course the infinitude of experiments cannot actually be carried out and has to be regarded as an unattainable ideal. When the definition is restated in a finite form the superficial appearance of objectivity becomes less convincing. This finite form is as follows. "The probability of the success of an experiment is p if, given $\varepsilon > 0$ and $\eta > 0$, there exists n_0 such that if $n_1 \geqslant n_0$ the proportion of successes in n trials differs from p by less than ε whenever $n_0 \leqslant n \leqslant n_1$, *with probability greater than* $1 - \eta$." Notice that the definition is now circular. η can be taken so small that the phrase "probability greater than $1 - \eta$" can be replaced for practical purposes by "certainty". This does not mean logical certainty but expresses an intense degree of belief. A supporter of the theory does not need to refer explicitly to degrees of belief. Instead, whenever he applies the above theorem he can make a definite prediction. But presumably he would not do this unless he did have an intense degree of belief.

(ii) *The "irregular collective" of von Mises.* (Axiomatic, objective, statistical, numerical.) The theory proposed by von Mises § is similar to the Venn limit but it avoids the difficulty of the definition being essentially circular by using the axiomatic method. Like any form of the frequency approach it can

---

† Venn, 1888. In essence this theory dates back at least as far as the seventeenth century. (See a quotation of Jacob Bernoulli's in Uspensky, 1937, 106.)

‡ The words "experiment" and "trial" will always be used in a very general sense.

§ R. von Mises, 1936 and 1945.

be applied only to experiments that can be conceived as one of a large class of similar experiments. von Mises deliberately restricts the theory of probability to such experiments. A central position in his theory is occupied by the " irregular collective " which will now be briefly described. Suppose that an infinite sequence of experiments is performed, and let " successes " be denoted by 1 and " failures " by 0. The results may thus be represented by a sequence of 0's and 1's such as 11010010 .... Such an infinite sequence is called an irregular collective if it has the following properties :—

(α) The proportion of 1's in the first n terms tends to a limit as $n \to \infty$. The limit may be called the probability, p, of success.

(β) More generally, if any subsequence is selected by means of a well-defined set of rules,† such that the question whether the mth term is selected is a function only of the previous $m - 1$ terms, then the proportion of 1's in this subsequence also tends to p. (In von Mises' formulation the " function " is a function of m only and does not depend on the first $m - 1$ elements of the collective. We prefer the present formulation since it expresses better " the impossibility of a gambling system ".)

Starting from these and similar assumptions it is possible to develop a detailed abstract theory. The method of applying this theory is to regard long sequences of trials as " approximately infinite ". This is equivalent to a judgment depending on degrees of belief and has the disadvantage of not being expressed in a precise form.

From the point of view of psychology any frequency approach has the advantage of being to some extent related to conditioned reflexes. For example, a dog will apparently regard a light signal as a probable indication of food provided that the signal has been followed by food in a high proportion of previous cases.

(iii) *The definition by equally probable cases, together with the " principle of*

† The question whether a sequence is an irregular collective depends on how the set of rules is defined. If the rules are defined in an unsuitable manner there would be no irregular collectives. For reasonable definitions we should expect irregular collectives to " exist " : but we should not want them to be mathematically constructible, since they would thereby lose an essential intuitive property of " randomness ". Some of the alleged disproofs of the existence of irregular collectives are based on the assumption that they are constructible. We add some further comments for the benefit of the reader who is familiar with point-set theory. Consider those sequences of 0's and 1's in which the proportion of 1's in the first n terms tends to p. Then it can presumably be proved, in the sense of Hausdorff fractional dimensions, that almost all of these sequences are irregular collectives, provided that the number of rules for determining subsequences is enumerable. This enumerability is a natural requirement, since there are at most an enumerable number of rules which can be laid down in a sentence of finite length using an unambiguous language. (For the theory of fractional dimensions see, for example, Hausdorff, *Math. Annalen*, **79** (1918), 157–79.) When $p = \frac{1}{2}$, Lebesgue measure is adequate. (See also Copeland, *Trans. Am. Math. Soc.*, **43** (1937), 333, and Wald, *Ergeb. math. Kolloqu. Hamburg*, **38** (1937), 38–72.)

insufficient reason" or "*the principle of cogent reason*". (Non-axiomatic, objective, non-statistical, numerical.) Suppose that when some hypothesis H is true there are exactly n equally probable "alternatives" and that a proposition E is necessarily true for m of them and necessarily false for the remaining ones. Then "the probability of E when H is assumed" is defined as m/n. In order to apply this definition it is necessary to be able to judge (or to know) that the various alternatives are equally probable. For example, if the hypothesis H is that we have a well-shuffled pack of playing cards and that the top card is drawn, then we may possibly judge that each of the 52 cards is equally likely to turn up. Therefore the probability that the card is either the ace or the two or the three of hearts is $\frac{3}{52}$. A method of judging that two cases are equally probable is by the "principle of insufficient reason", i.e. the two cases are equally probable if there is no conceivable reason to expect one rather than the other. Such a judgment is liable to be made when there is some sort of symmetry, and the principle invoked is then more accurately described as "the principle of cogent reason".† But there will always be some difference between the two cases in any practical example, and it will be necessary to decide that the differences are unimportant. For example, it might be argued that a card with more print on it is likely to be slightly heavier and that this upsets the symmetry. The rules for deciding when such departures from symmetry are important have never been clearly stated.

Several probability experiments have been made with the intention of showing that the theories (i) and (iii) give the same results when they are both applicable. Such experiments have usually given good results, but they cannot *prove* anything.

The conflict between definitions (i) and (iii) is an old one. Those who define probability in terms of equally probable cases say that the frequency with which things happen cannot be fundamental since it can only modify previously known probabilities. Their opponents reply that these probabilities could themselves have been based only on previous experience in any real problem (since complete symmetry is unobtainable). They may also say that the principle of cogent reason is itself a generalisation from experience.

On the whole the frequency approach seems to be more popular among physicists. But E. C. Kemble (1942) considers that it is inadequate for problems occurring in statistical mechanics, though justifiable in some circumstances.

(iv) *Jeffreys' theory*. (Axiomatic, objective,‡ non-statistical, numerical (essentially).) This is similar to theory (iii) but it uses the axiomatic method. No definite distinction is drawn between the axioms and the rules of application of the theory. Jeffreys considers that for a given proposition or "event" E

† See A. Fisher, 1922.
‡ See classification (*b*) on page 6 and a footnote on page 2.

and for given hypotheses H, there is only one reasonable degree of belief, and that any two such degrees of belief are comparable. He obtains a numerical theory and provides suggestions (rather than axioms) for obtaining the numerical probability for a number of problems. In all these problems it is necessary to apply the principle of cogent reason, and therefore the criticism of definition (iii) still applies. A comprehensive account is given by Jeffreys (1939).†

(v) *The definition by point-set theory.* (Axiomatic, numerical.) It is possible to represent the results of most scientific experiments by a finite set of measurements, i.e. by a point in a finite-dimensional space. The probability that the result of the experiment will be a point belonging to a particular set in this space can be taken as the " measure " of this set, where the measure may be interpreted in the Lebesgue sense, or in any of a number of other senses. In this way it is possible to establish an abstract theory of probability. This method was first used by Kolmogoroff (1933). (See also Cramér (1937).) The appropriate measure has to be decided upon before the theory can be applied, and this choice of measure is equivalent to a judgment of equally probable cases. This point is made by Jeffreys (1939), 302. If Lebesgue measure is invariably used the theory becomes self-contradictory.‡ Whether the method is an axiomatic form of method (iii) depends on the rules given for its application.

(vi) *Probability defined as a " proportion of possible alternatives ".*§ (Non-axiomatic, objective, neither statistical nor dependent upon degrees of belief, numerical.) This definition is ambiguous since there is no unique way of defining the " possible alternatives ", and different results are obtained according to the method used. Suppose, for example, that it is known that of a set of three billiard balls the two white ones are kept in one drawer and the red ball in another drawer. One of the drawers is opened and a ball is selected. What is the probability that it is the red one ? It might be said that there are exactly three alternatives since there are three balls, so that the probability is $\frac{1}{3}$. Or it might be said that there are two alternatives since there are two drawers that can be opened, and the drawer that is opened determines the colour of the ball selected, so that it is unnecessary to split the alternatives up any further. This would make the probability $\frac{1}{2}$. (Cf. Jeffreys (1939), 301.)

---

† It may be mentioned in passing that what Jeffreys calls " convention 2 " really amounts to an extra assumption. For it can be used to prove that a " perfect " seven-sided die has less probability of giving a 6 than an ordinary die—a result not otherwise deducible from his axioms. The trouble can be removed by replacing the equalities in his axiom 4 by inequalities.

‡ The invariable use of Lebesgue measure would be equivalent to an uncritical use of " Bayes' postulate ". (See 5.3.)

§ This is called the " finite frequency theory " by Bertrand Russell, *loc. cit.*, 368. W. Kneale, in *Probability and Induction* (Oxford, 1949), expresses the opinion that it is only in terms of some such theory that objective probabilities can be considered to exist.

If the number of alternatives is infinite the position is even worse, since it is meaningless to talk about a proportion of an infinite number of things, unless a definite limiting process is specified. The definition might be made applicable if a set of rules could be given for deciding on a unique set of possible alternatives for every example. But such a set of rules seems unlikely ever to be produced.

(vii) *Ramsey's theory*.† (Axiomatic, not entirely objective, neither statistical nor dependent only on degrees of belief, numerical.) In this theory *expected benefit* is taken as a more fundamental idea than degrees of belief. Degrees of belief are defined in terms of expected benefits instead of the other way round as in most theories. (In any case a scale of values or "utilities" must be assumed.) It is not clear whether Ramsey's method is always justifiable in the applications to purely scientific problems. At least it suggests the possibility of extending our "body of beliefs" so as to include judgments of the type that one expected benefit is greater than another one.

(viii) *Koopman's theory*.‡ (Axiomatic, not objective, non-statistical, non-numerical at first.) The essence of this method is given by its classification. It is not supposed to be applicable without using what we have called a "body of beliefs". Koopman deduces a numerical theory for a class of problems, from a more general non-numerical theory. He has been much influenced by the work of J. M. Keynes (1921) whose theory may be classified thus: axiomatic, objective, non-statistical, non-numerical (in general). Keynes in his turn was influenced by W. E. Johnson's lectures and conversations. In 1931 Keynes admitted § that he no longer adhered to an objective theory. But it is possible to salvage the formal apparatus of his theory.

(ix) *Orthodox statistical theories*.‖ (Axiomatic, objective, statistical, numerical.) Any theory with the classification shown may be called an orthodox statistical theory. Hence this class of theories includes von Mises' theory (ii) as a special case. It also includes theory (v) if that theory is interpreted in terms of what happens "in the long run". There is a considerable choice in the form of the axioms of an orthodox statistical theory, and it is not at all necessary that they should depend on ideas akin to that of the irregular collective. But most of what we shall say would apply equally well to theory (ii).

Any orthodox statistical theory is a scientific theory in almost exactly the same sense as geometry: there is a rigorous mathematical theory and a non-rigorous technique for applying the theory. Degrees of belief are not a part of the theory, but they are used when the theory is applied, just as they are used

---

† F. P. Ramsey, 1931, Chapters 7 and 8.
‡ See Koopman, 1940.
§ *Essays in Biography* (London, 1933), 300.
‖ See, for example, Bartlett, 1940, or Reichenbach, 1932.

when any other scientific theory is applied. A probability in the theory is regarded as something objective, like the distance between two points.

Bartlett's view is that it is valuable to have *two separate theories*, one for degrees of belief and the other for objective probabilities.† My view is that if a single theory covers both the objective and subjective aspects so much the better. Thus, while admitting the importance of the practical distinction between objective probabilities and reasonable degrees of belief, I consider that each objective probability is at the same time the only reasonable degree of belief. (This is discussed in more detail in 4.9.) The advantage of two separate theories is to emphasise the distinction between the objective and subjective aspects. But I find it philosophically more satisfying and more economical to have a single theory. I consider that in the last resort one must define one's concepts in terms of one's subjective experiences. (This does not necessitate philosophical solipsism.) The opposite view is that degrees of belief can be interpreted only by the methods of experimental psychology.

The orthodox statistical theories do not deal with the problem of scientific induction, but rather they need to be justified by induction. This problem of induction is a problem of what to believe, and for it a theory of degrees of belief is appropriate.

An important property of the theories (i) to (ix) is that they cannot be applied without the use of judgment, so that really none of them is objective in any absolute sense. An advantage of Koopman's theory is that it is made quite clear what sort of judgments are to be used. The theory in the present book is similar to Koopman's, but the axioms and the development of the abstract theory are simpler. In order to achieve this simplicity some sacrifice has to be made. The sacrifice is that it is assumed in the axioms that probabilities correspond to numbers; but this assumption is not completely used in the applications. The theory adopted has the classification: *axiomatic, not necessarily objective* (*though objectivity is not ruled out*), *non-statistical on the whole, not entirely numerical*.

For the benefit of those who are familiar with Jeffreys' theory, a few remarks showing the relation between his theory and ours will not be out of place. Our theory resembles that of Jeffreys in the use of the symbol $P(E \mid H)$. This symbol is, however, given a double interpretation, only one of which is numerical. (See 4.1.) The following are the main differences between the two theories :—

(a) Our emphasis is on the *comparisons* between beliefs, thereby avoiding the necessity of making judgments of exactly equal intensities of belief.

(b) The beliefs in any problem are regarded as depending on the individual concerned.

† This dualistic view is shared by Nagel, Carnap and Koopman. See, for example, the excellent reviews by Koopman in *Math. Rev.*, 7 (1946), 186–93.

(c) There is a splitting into axioms, rules and suggestions, as explained in Chapter 4. This shows clearly what parts of the theory depend on pure mathematics and logic only and what parts can be varied according to taste. Given the primitive notion of a comparison between degrees of belief, the rules of application are absolutely precise. This is not true of the " suggestions ", but these are not an essential part of the theory.

(d) There is no dependence on the principle of cogent reason. Any apparent application of this principle is in reality a subjective judgment which is made without *direct* reference to any central authority. Similarly there will be subjective judgments that may appear to be concessions to the frequency definition, but which are really a result of a familiarity with a theorem corresponding to this definition. Some such mixture of the two classical approaches is the way in which most people have *used* probability for the last 300 years. It is therefore claimed that our theory is more closely related to practice than are most theories of probability.

CHAPTER 2

THE ORIGIN OF THE AXIOMS

2.1 The purpose of this chapter is to show that the axioms stated in Chapter 3 are not chosen in a haphazard manner. *The arguments will not be very rigorous.* The plan is to take theory (iii) of Section **1.4**, the "definition" by equally probable cases, and to apply it to a class of problems in which it may well be judged that various events are equally probable. Such problems are provided by some idealised games of chance. Our method is thus closely related to the historical development.

It is equally possible to provide a rough justification by using theories (i) or (v). The method chosen has the advantage of avoiding infinite sequences of trials and advanced mathematics. The main result of the chapter will be to suggest two axioms, known as the laws of addition and multiplication. With theory (i) both laws would be simple theorems; by contrast, when probabilities are interpreted as degrees of belief, attempts have been made to show that these laws are mere conventions. (See, for example, Schrödinger,† 1947. But see also the footnote in **1.4** (iv) concerning Jeffreys' "convention 2".)

Further remarks about the *a priori* justification of the axioms will be found in **4.1A**.

Before carrying out the main plan of the chapter we shall consider how far it is possible to go by relying only on what is intuitively "obvious".

2.2 Two "obvious" axioms

Let E_1, H_1, E_2 etc. be various propositions, and for short write p_1 for $P_\mathcal{B}(E_1 \mid H_1)$, p_2 for $P_\mathcal{B}(E_2 \mid H_2)$, etc. Here p_1 and p_2 do not represent numbers, but are simply symbols for degrees of belief. Now it may happen that one of the comparisons belonging to $\mathcal{B}$ is that p_1 is greater than p_2, i.e. that the belief in E_1, when H_1 is assumed, is more intense than the belief in E_2 when H_2 is assumed. In this case we may say for short that $\mathcal{B}$ includes "$p_1 > p_2$". Equally $\mathcal{B}$ may include "$p_2 > p_1$". On the other hand, p_1 and p_2 may not be comparable in $\mathcal{B}$.

There are now two axioms that are virtually forced upon us. The first is that "$p_1 > p_2$" and "$p_2 > p_1$" are not both parts of $\mathcal{B}$, or if they are then $\mathcal{B}$

---

† Schrödinger's argument depends largely on the very natural assumption that the probability of the disjunction of a number of mutually exclusive propositions is a function of the separate probabilities. (See also Appendix III.)

must be regarded as unreasonable.† The second is the "transitive" property of the relation " $>$ " : if $p_1 > p_2$ and $p_2 > p_3$ are both parts of $\mathcal{B}$, then $p_1 > p_3$ may be added to $\mathcal{B}$ (if it is not already included).‡ Like the first axiom this may lead to a contradiction.

These two axioms are notable in virtue of their obviousness. It does not seem to be possible to develop a useful axiomatic theory of probability without using some axioms that are less obvious than these two. In this respect probability differs from classical formal logic.

In the next section we shall talk about probabilities that are judged to be "equal" (i.e. equally intense). This is not meant to imply that such judgments are necessarily possible in practice (except between logical certainties and impossibilities). It is merely part of the plan mentioned at the beginning of the chapter.

For the rest of this chapter the word "probability" will be used in the sense of the "equally-probable-cases" definition.

2.3 Definition of numerical probability by judgment of equally probable alternatives

Two propositions A and A' are said to be "mutually exclusive given H" or "incompatible given H" if $A.A'$ is necessarily false on the assumption that H is true. A number of propositions are said to be "exhaustive given H" if one of them must be true when H is true.

Let $A_1, A_2, \ldots, A_n$ be n propositions that are mutually exclusive and exhaustive given H. Suppose further that they are judged to be equally probable (given H). This judgment is of course part of the body of beliefs, $\mathcal{B}$. Let

$$E = A_1 \vee A_2 \vee \ldots \vee A_m \quad (0 \leqslant m \leqslant n).$$

Then we define $P_{\mathcal{B}}(E \mid H)$ or $P(E \mid H)$ as m/n. In words, "the probability of E given H is the proportion of equally probable alternatives in which E is true given H". Essentially this is a restatement of the definition of **1.4** (iii).

The possibilities $m = 0$ and $m = n$ correspond to propositions E which are respectively impossible or certain given H. In fact, if E is any proposition which is impossible or certain given H, we can express E in the above form, and thus show that its probability is 0 or 1. For we may take $n = 1$, $A_1 = H$, $m = 0$, or $m = n = 1$, $A_1 = H = E$ respectively.

There are two immediate criticisms of the definition. The first is that

---

† This is essentially a repetition of a point made in Section 1.3.

‡ If $\mathcal{B}$ is enlarged in this way so as to become "transitive", then it may be regarded as a "partially ordered system". See G. Birkhoff, *Lattice theory* (Amer. Math. Soc., 1940), chapter 1. Partial ordering is an essential part of Keynes's theory. Jeffreys, in the preface to the second edition of *Probability* (1948), erroneously asserts that Keynes withdrew the suggestion of partial ordering in his *Essays in biography*. (See 1.4, viii.)

there may be no way in general of expressing any given proposition E in the required form. The second is that there may be more than one way, and the corresponding values of $P(E \mid H)$ may not be equal. The answer to the first criticism is that we are at present restricting our attention to those cases in which the alternatives $A_1, A_2, \ldots, A_n$ can be found. As regards the second criticism, we propose to assume, merely as a plausible hypothesis, that $P_\mathfrak{B}(E \mid H)$ cannot have two different values, provided that $\mathfrak{B}$ is sound. This is of course not an additional assumption if the A's are unique.

It is impossible to prove that the definition is in any sense the right one. It is a simple and natural method of correlating numbers with degrees of belief in a class of ideal cases, and it is very nearly obvious that it has the effect of assigning larger numbers to more intense rational degrees of belief. Any monotonic function of m/n could be chosen instead and would have the same property, but the effect would be to complicate the theory unnecessarily. This possibility of choosing an arbitrary monotonic function is related to the question of whether the definition is only a convention.

2.4 *Example*

In order to be convinced that the definition just given has any significance it is advisable to consider an example.

Imagine an ordinary pack of playing cards that has been well shuffled and placed face-downwards on the table. There is no special reason for supposing that, say, the three of hearts is more likely to be the top card than the seven of spades. If there is such a reason for some real pack of cards we could imagine the pack to be replaced by a " perfect " pack in which there is no such reason. It is difficult to believe that this would force us into an untenable position. Suppose then that we are dealing with such a perfect pack. The object here is not to obtain approximations for the probabilities in the case of a real pack, but merely to show that there are ideal circumstances in which the definition of **2.3** can be applied.†

For simplicity suppose that the cards are numbered from 1 to 52. Let A_r be the proposition that the top card is number r. Let H be a physical description of how the experiment is carried out. The description must not be too complete, since the very notion of probability depends on an assumption of partial ignorance. (We are ignoring here the insoluble problem of " determinism " versus " indeterminism ".) As it happens it is usually impracticable to provide a description that is so complete as to make a precise prediction

† If the present chapter had been based on the frequency definition it would also have been necessary to consider idealised problems, since this definition involves infinite sequences of experiments. Which idealisation is regarded as more natural is a matter of taste.

possible. H may be thought of roughly as "the pack is very well shuffled". Let $\mathfrak{B}$ consist of the assertion that $A_1, A_2, \ldots, A_{52}$, are all equally probable given H.

It can now be stated, for example, that the probability that the top card is black (given H) is $\frac{1}{2}$.

The reader would have no difficulty in inventing other examples, using perfect coins, dice or roulette wheels, in which the natural numbers of alternatives are 2, 6 and 37 respectively.

2.5 The law of addition of probabilities

Suppose that with the assumptions of Section **2.3**,
$$E = A_1 \vee A_2 \vee \ldots \vee A_m \quad (0 \leqslant m \leqslant n),$$
$$F = A_{m+1} \vee A_{m+2} \vee \ldots \vee A_{m+r} \quad (m + r \leqslant n).$$
Clearly E and F are mutually exclusive and $P(E \mid H) = m/n$, $P(F \mid H) = r/n$. Moreover
$$E \vee F = A_1 \vee A_2 \vee \ldots \vee A_m \vee A_{m+1} \vee \ldots \vee A_{m+r},$$
so that $P(E \vee F \mid H) = (m + r)/n$, i.e.
$$P(E \vee F \mid H) = P(E \mid H) + P(F \mid H) \quad . \quad . \quad . \quad (1)$$
This is called the law of addition of probabilities. *It is essential that E and F should be mutually exclusive (given H).*

There is no difficulty in extending equation (1) to the disjunction of more than two mutually exclusive propositions.

Exercise. When is it legitimate to put $E = F$ in equation (1)?

Example. Consider the well-shuffled pack of cards already mentioned. What is the probability that the top card will be either a diamond or the ace of spades? These two events are mutually exclusive and have probabilities $\frac{1}{4}$ and $\frac{1}{52}$ respectively. Hence the required probability is the sum of these numbers, i.e. $\frac{7}{26}$. This may be at once verified from the original definition. On the other hand, the probability that the top card will be a spade or an ace is not $\frac{1}{4} + \frac{1}{13}$, for this time the events are not mutually exclusive.

2.6 The law of multiplication of probabilities

Let E and F be any two propositions that are expressible as a disjunction of the A's, where the A's and H are defined as before. Without loss of generality it may be supposed that
$$E = A_1 \vee A_2 \vee \ldots \vee A_m \quad (0 \leqslant m \leqslant n),$$
$$F = A_1 \vee A_2 \vee \ldots \vee A_r \vee A_{m+1} \vee A_{m+2} \vee \ldots \vee A_{m+s} \quad (r \leqslant m, m + s \leqslant n).$$
(E and F can be put in this form by renumbering the A's if necessary.) Then
$$E.F = A_1 \vee A_2 \vee \ldots \vee A_r.$$
Therefore $P(E.F \mid H) = r/n$. Moreover $P(E \mid H) = m/n$ and in order to reach our objective, namely equations (2) below, it remains to prove that

ORIGIN OF THE AXIOMS

$P(F \mid E.H) = r/m$. Now $A_1, A_2, \ldots, A_m$ are equally probable given H, and if in addition we know that E is true, i.e. that one of $A_1, A_2, \ldots, A_m$ is true, then it is *very natural* to assume that $A_1, A_2, \ldots, A_m$ remain equally probable since the additional information is symmetrical with regard to these propositions. In fact we shall suppose that part of $\mathfrak{B}$ is that $A_1, A_2, \ldots, A_m$ are equally probable given $E.H$. Now $A_1, A_2, \ldots, A_m$ are mutually exclusive and exhaustive given $E.H$. Therefore $P_\mathfrak{B}(F \mid E.H) = r/m$, as asserted. Thus

$$P(E.F \mid H) = P(E \mid H).P(F \mid E.H) \qquad . \qquad . \qquad (2)$$

This is the law of multiplication of probabilities.† If H is taken for granted (a practice that is apt to be misleading) we could write ‡ for short $P(E.F) = P(E).P(F \mid E)$, or, in words, "the probability of the conjunction of two propositions is the product of the probability of the first with that of the second *given the first*". It may happen that E and F are "independent" § given H. In this particular case the equation (2) takes the simpler form

$$P(E.F \mid H) = P(E \mid H).P(F \mid H) \qquad . \qquad . \qquad (2A)$$

Exercise. When is it legitimate to put $E = F$ in this formula?

2.7 Example

Two "perfect" dice are thrown. What is the probability of obtaining two sixes?

Let us suppose that a beginner has a body of beliefs which includes the following judgments.

(*a*) The six possible results of the first throw are equally probable.

(*b*) The 36 possible results of the pair of throws are equally probable.

(*c*) The probability of a 6 on the second throw is increased (or decreased) by a knowledge that the first throw resulted in a 6.

The judgment (*b*) gives $\frac{1}{36}$ as the answer to the problem. On the other hand the judgments (*a*) and (*c*), together with the law of multiplication of probabilities, give a result that is either greater or less than $\frac{1}{36}$. Hence the body of beliefs is inconsistent with a formal use of the law of multiplication.

2.8 Continuous probabilities

In the definition of **2.3** a probability was necessarily measured by a rational number. Such probabilities may be sufficient for all applications to the real world, but they are not sufficient for some types of idealised problems. As a simple example suppose that a decimal is chosen between 0 and 1 in such a

---

† The above proofs of the addition and multiplication laws may easily be generalised to propositions E and F which do not imply H.

‡ But see the second paragraph of **3.2**.

§ i.e. if one is assumed the probability of the other is unaffected.

2.8 PROBABILITY AND WEIGHING OF EVIDENCE

way that each of its digits is judged to have an equal and independent † probability of being one of the numbers 0, 1, 2, . . ., 9. An infinite number of choices must be imagined. Within the framework of any standard theory of probability, this is equivalent to the selection of a point P on a line AB of unit length in such a way that for *each* fixed length the point is equally likely to lie in any interval of that length. (In these circumstances P is said to "have a uniform distribution of probability over AB".) It is then easily proved that if CD is a sub-interval of positive rational length then the probability that P will lie in CD is equal to the length of CD. It is natural to suppose that this applies even if CD is irrational.‡ This shows that it may be convenient to allow irrational numbers to represent probabilities. Another peculiarity of this problem is that the probability of P being exactly at the given point D is zero. (This is the degenerate case in which C and D coincide.) But it is not logically impossible that this should happen. We therefore introduce a new definition. If $P(E \mid H) = 0$ we say that E is *almost impossible* given H. Impossibility implies almost impossibility but not conversely. *Almost certain* can be defined in a similar way.§

Ideas of this sort occur frequently in problems in which probability depends on position in space or time. In practice we can measure space and time only to a finite number of places of decimals, but it is often simpler to imagine that the measurements are capable of being equal to any real number of units. If we were satisfied to deal only with entirely practical problems it would hardly be necessary to distinguish between "impossible" and "almost impossible".

There are other types of problems in which these ideas are convenient, namely when infinite sequences of trials are imagined. Some important examples will occur in the sequel.

---

† i.e. not depending on a knowledge of any selection of the other digits.
‡ This can be formally proved by assuming axiom 1 and theorem 13 of Chapter 3.
§ These definitions are suggested by standard terminology in the theory of "measure", and they have been used by previous writers.

CHAPTER 3

THE ABSTRACT THEORY

3.1 The axioms

The notation of **1.1** will be used, and it will be taken that the propositions E, H etc. never involve probabilities or beliefs. A symbol H^* is introduced which is supposed to represent all the usual basic assumptions of logic and pure mathematics. (It is conceivable that H^* is not expressible in a finite number of words, but it will be regarded as a proposition.) Any proposition that is implied by H^* is called "logically true" or "certain" and its negation is called "logically false" or "impossible". A logically true proposition is also known as an "analytic proposition". There is a difference of opinion as to the meaning of a "proposition", as to what should be included in H^* and as to the meaning of implication by H^*. No attempt will be made here to decide these questions: a different theory of probability will correspond to each possible answer. For any two propositions E and F, "E implies F" means that $\bar{E} \vee F$ is a logically true proposition.

Symbols of the form "$P_\mathfrak{B}(E \mid H)$" = "$P(E \mid H)$" are introduced. They are read "the probability of E given H (and assuming $\mathfrak{B}$)" and are otherwise undefined.† Within the abstract theory the word "probability" should not be interpreted in terms of beliefs.

The axioms are numbered A1 to A6.

A1 $P(E \mid H)$ is a non-negative real number.
A2 If $P(E.F \mid H) = 0$, then $P(E \vee F \mid H) = P(E \mid H) + P(F \mid H)$.
A3 $P(E.F \mid H) = P(E \mid H).P(F \mid E.H)$.
A4 If E and F are logically equivalent (i.e. if they imply one another) then
    $P(E \mid H) = P(F \mid H)$ and $P(H \mid E) = P(H \mid F)$ for any H.
A5 $P(H^* \mid H^*) \neq 0$.
A6 $P(E^* \mid H^*) = 0$ for some proposition E^*.

Remarks

(i) When the definition by equally probable cases can be applied in order to define (as a rational number) all the probabilities that occur, then, as in Chapter 2, we can deduce axioms A2 and A3 together with $0 \leqslant P(E \mid H) \leqslant 1$, $P(H^* \mid H^*) = 1$ and $P(\bar{H}^* \mid H^*) = 0$. The last three deductions clearly imply A1, A5 and A6, which are therefore preferable on grounds of economy. Finally

† Some variations of language will occur. For example, the words "given" and "assuming" may be interchanged.

3.1 PROBABILITY AND WEIGHING OF EVIDENCE

A4 is suggested directly by the interpretation of probability as a reasonable degree of belief.

The axioms are formally suggested but are not proved by Chapters 1 and 2. There are perhaps less restrictions than before on the propositions E and H, and the question of self-consistency is therefore more pressing now. This question will be discussed in **3.4** and **4.1A**.

(ii) A4 enables us to write, for example, $P(E \mid H.H^*) = P(E \mid H)$. It would be wrong to regard A4 as entirely obvious when interpreted in terms of reasonable beliefs. A possible modification of this axiom will be considered in **4.13**.

(iii) The "obvious" axioms of **2.2** are automatically satisfied in a sense to be described.

It will be seen in the next chapter that full use is never made of the assumption that the probabilities of the abstract theory are numbers. But the assumption has the great merit of simplicity. If one numerical probability is greater than another one, say $P(E \mid H) > P(E' \mid H')$, then in the applications this is interpreted in the natural way in terms of reasonable beliefs. It is in this sense that the "obvious" axioms are satisfied. But this interpretation in terms of beliefs does not belong to the *abstract* theory and further discussion of it is postponed until the next chapter.

(iv) Chapter 2 suggests that logical certainty and impossibility should be represented by probabilities of 1 and 0 respectively. Accordingly it might have been assumed that

(a) if H implies E then $P(E \mid H) = 1$,

(b) if H implies $\bar{E}$ then $P(E \mid H) = 0$.

But these two axioms would lead to trouble. For they give $P(E \mid E.\bar{E}) = 0$ and also $P(E \mid E.\bar{E}) = 1$.† It may be possible to avoid this contradiction by insisting that in the expression $P(E \mid H)$ the proposition H should never be self-contradictory. A more formal method of avoiding the difficulty is provided by the adoption of A5 and A6.

(v) In all this work the symbol $\mathfrak{B}$ is taken for granted. It may be thought of as a set of inequalities and equalities between (numerical) probabilities, but its exact form is unimportant as far as this chapter is concerned.

(vi) The development of the abstract theory must follow the rules of ordinary logic and pure mathematics. Hence we could, at this stage, hardly allow the propositions E, F, H, etc. to involve probabilities. This is the reason for the convention at the beginning of the chapter. To what extent this restriction may be relaxed is an interesting question. If it were entirely relaxed it would enable us to write $P(E \mid H.\mathfrak{B})$ instead of $P_\mathfrak{B}(E \mid H)$, and this would at once suggest an extension of the axioms. The resulting theory would have some

† The proposition $E.\bar{E}$ implies both E and $\bar{E}$.

THE ABSTRACT THEORY 3.2

convenience, but it would also be confusing and might even be self-contradictory. The question is mentioned again in **4.9**.

(vii) The practical significance of the axioms will not appear until Chapter 4. The whole of the abstract theory can be deduced from the axioms without relying at all on any interpretation of probability.

(viii) The choice of axioms is related to the historical background of the subject, but no attempt will be made to trace this aspect of the matter. Other sets of axioms can be used instead.† One such set will be given in **3.4**.

(ix) The axioms are equally strongly suggested by a point-set approach. (Cf. **1.4** (v).) For example, suppose that E is the proposition asserting that the result of an experiment consists of a set of n real numbers, which, regarded as a point in n-dimensional space, belongs to a certain measurable set of points $\mathfrak{G}$. Define $P(E)$ as the measure of the set $\mathfrak{G}$ divided by the measure of the whole space, assuming the denominator to be finite. Define $P(E \mid H)$ as $P(E.H)/P(H)$ if $P(H) \neq 0$. Let the set corresponding to H^* be the whole space. All the axioms can be proved with these definitions and restrictions. This lends support to the self-consistency of the axioms. In some idealised problems it may be convenient to allow the whole space to have infinite measure and to define $P(E)$ simply as the measure of $\mathfrak{G}$. This leads to a slightly different abstract theory in which certainty is represented by infinity instead of by unity. (Cf. Jeffreys (1939), 21 and 114.)

3.2 Definitions

The definitions, like the axioms, are suggested in part by Chapter 2.

The symbol ‡ $P(E)$ may be written as an abbreviation for $P(E \mid H^*)$ and may be read "the probability of E". If $P(E) = 0$, E is *almost impossible* and if $P(E) = 1$, E is *almost certain*.

If $P(E.F \mid H) = 0$, E and F are *almost mutually exclusive given* H. If $P(E.F) = 0$, E and F are *almost mutually exclusive*. If every pair of E_1, E_2, E_3, ... are almost mutually exclusive (given H), then E_1, E_2, E_3, ... are *almost mutually exclusive* (given H).

If $P(F \mid E.H) = P(F \mid H)$, then F is *independent* § of E given H. If $P(F \mid E) = P(F)$, F is *independent of* E. If each of a finite set of propositions E_1, E_2, E_3, ... is independent of the conjunction of any number of the rest (given H) then E_1, E_2, E_3, ... *are independent* (given H).

The object of these definitions is to make the statements of the theorems

† See, for example, C. D. Broad, "Hr. von Wright on the logic of induction (II)", *Mind*, **53**, 1944, 97–119.

‡ This should not be confused with the "misleading" notation of **2.6**, **5.1** and elsewhere.

§ It might have been better to call this condition "almost independence" to distinguish it from other meanings of the word "independence". But the above definition is unlikely to cause confusion.

21

3.3   PROBABILITY AND WEIGHING OF EVIDENCE

more concrete and therefore easier to grasp and to remember. But the phrase "E is almost impossible (given H)" will usually be avoided because its systematic use would be rather monotonous. The equation "$P(E \mid H) = 0$" will be written instead, and it is left to the reader to interpret this in accordance with the definition of almost-impossibility if he wishes to do so. Similarly the phrase "almost certain" will often be avoided.

3.3 Theorems

The first eight theorems depend only on axioms A1 to A4.

T1   If F is independent of E given H, then
$$P(E.F \mid H) = P(E \mid H).P(F \mid H). \qquad (1)$$
This is an important special case of A3.

T1A   If either $P(E \mid H) = 0$ or $P(F \mid H) = 0$ then the equation (1) holds without the assumption of independence. (Proof by A1 and A3.)

T2   If $E_1, E_2, \ldots, E_n$ are almost mutually exclusive given H, then
$$P(E_1 \vee E_2 \vee \ldots \vee E_n \mid H) = P(E_1 \mid H) + P(E_2 \mid H) + \ldots + P(E_n \mid H),$$
and the two propositions $E_1 \vee E_2 \vee \ldots \vee E_{n-1}$ and E_n are almost mutually exclusive.

The two parts can be proved simultaneously by induction. The theorem is true when $n = 2$, by A2. Suppose it is true when $n = m$. Then it is sufficient to show that $E_1 \vee E_2 \vee \ldots \vee E_m$ and E_{m+1} are almost mutually exclusive given H. Now if i and j are less than $m + 1$,
$$P\{(E_i.E_{m+1}).(E_j.E_{m+1}) \mid H\} = P(E_i.E_j.E_{m+1} \mid H), \text{ by A4,}$$
$$= P(E_j.E_{m+1} \mid H).P(E_i \mid E_j.E_{m+1}.H), \text{ by A3,}$$
$$= 0,$$
since E_j and E_{m+1} are almost mutually exclusive given H. Therefore
$$P\{(E_1.E_{m+1}) \vee (E_2.E_{m+1}) \vee \ldots \vee (E_m.E_{m+1}) \mid H\}$$
$$= P(E_1.E_{m+1} \mid H) + P(E_2.E_{m+1} \mid H) + \ldots + P(E_m.E_{m+1} \mid H),$$
by the inductive hypothesis, and each term of this sum is 0. Thus by A4,
$$P\{(E_1 \vee E_2 \vee \ldots \vee E_m).E_{m+1} \mid H\} = 0$$
as required.

It is impossible to prove a result corresponding to T2, for an infinite number of propositions. If such a result is required it must be assumed as an axiom.

If E is the disjunction of an enumerable set of almost mutually exclusive propositions $E_1, E_2, E_3, \ldots$, it is easy to prove, using T13, that
$$P(E \mid H) \geqslant P(E_1 \mid H) + P(E_2 \mid H) + \ldots, \text{ if } P(H) \neq 0.$$
The additional axiom would replace the inequality by an equality. Such a new axiom is not essential but it has applications in some types of idealised problems. As a matter of fact it is not required if it is assumed that
$$P(E_n \vee E_{n+1} \vee E_{n+2} \vee \ldots \mid H) \to 0 \text{ as } n \to \infty.$$
This assumption would be a natural one in any application that is likely to arise.

THE ABSTRACT THEORY   3.3

The additional axiom may be called the *axiom of complete additivity*.† With its help it can be proved for example that for any infinite sequence of propositions F_1, F_2, F_3, . . .,

and
$$P(F_1 \vee F_2 \vee F_3 \vee \ldots) = \lim_n P(F_1 \vee F_2 \vee \ldots \vee F_n),$$
$$P(F_1.F_2.F_3.\ldots) = \lim_n P(F_1.F_2.\ldots F_n).$$

The axiom of complete additivity corresponds to a similar property of point-sets that are measurable in the Lebesgue sense. Hence it could be introduced without serious risk of inconsistency; but in the present book it will never be used except as a mathematical convenience, and with the understanding that its use could be avoided.

T3 If F is independent of E given H, then E is independent of F given H, assuming that $P(F \mid H) \neq 0$.

PROOF. $P(E.F \mid H) = P(E \mid H).P(F \mid H)$ by T1. But
$P(E.F \mid H) = P(F \mid H).P(E \mid F.H)$ by A3.
Therefore by equating these two values of $P(E.F \mid H)$ we obtain
$$P(E \mid F.H) = P(E \mid H) \text{ if } P(F \mid H) \neq 0.$$
This theorem may be stated: "If F is independent of E and F is not almost impossible, then E and F are independent (given H in each case)." (See the last definition of **3.2**.)

T4 For any finite set of propositions E_1, E_2, E_3, . . .

$P(E_1.E_2.E_3.\ldots \mid H) = P(E_1 \mid H).P(E_2 \mid E_1.H).P(E_3 \mid E_1.E_2.H) \ldots$
(Proof by induction from A3.)

T5 If the finite set of propositions E_1, E_2, E_3, . . . are independent given H, then
$$P(E_1.E_2.E_3 \ldots \mid H) = P(E_1 \mid H).P(E_2 \mid H).P(E_3 \mid H) \ldots$$
This is a special case of T4 or may be proved by induction from T1.

Example. Suppose that E and F are independent, F and G are independent, and G and E are independent (given H in each case). Then it does *not* follow that
$$P(E.F.G \mid H) = P(E \mid H).P(F \mid H).P(G \mid H).$$
To see this *intuitively* let the propositions E, F, G be defined as follows:—

E: Smith has green eyes.

F: The next man you meet will be Smith.

G: The next man you meet will have green eyes.

No attempt will be made to specify H and $\mathcal{B}$.

In this example $E.F.G = F.G$ so that
$$P(E.F.G \mid H) = P(F.G \mid H)$$
$$= P(F \mid H).P(G \mid H).$$
This is not equal to $P(E \mid H).P(F \mid H).P(G \mid H)$ in general.

† Cf. Fréchet, 1937, 22; Cramér, 1937, 9; Kolmogoroff, 1933, 13.

3.3   PROBABILITY AND WEIGHING OF EVIDENCE

T5A The formula of T5 applies if any of $P(E_1 \mid H)$, $P(E_2 \mid H)$, ... vanishes, without the assumption of independence. (Cf. T1A.)

T6 *Bayes' theorem.* If E is a variable proposition and F and H are fixed, then
$$\frac{P(E \mid F.H)}{P(E \mid H)} \text{ is proportional to } P(F \mid E.H),$$
assuming that $P(E \mid H) \neq 0$ and that $P(F \mid H) \neq 0$.

Proof. $P(E \mid H).P(F \mid E.H) = P(E.F \mid H)$
$$= P(F \mid H).P(E \mid F.H).$$
Therefore
$$\frac{P(E \mid F.H)}{P(E \mid H)} = \frac{P(F \mid E.H)}{P(F \mid H)},$$
assuming that $P(E \mid H) \neq 0$, $P(F \mid H) \neq 0$. The result follows at once.

There has been a great deal of dispute about the validity of this theorem and about its applicability. If we think of the various E's as being a set of possible theories (or hypotheses) and F as a proposition describing the results of some experiments, then we may regard $P(E \mid H)$ as the *initial* or *prior probability* of the theory E and $P(E \mid F.H)$ as its *final* or *posterior probability*.† The theorem may then be stated: "The ratio of the final to the initial probability of a theory ‡ is proportional to the probability (given E and H) of the observed results of experiments." More will be said about Bayes' theorem in other chapters.

Before going on to theorem 7 the reader should consider what happens to theorems 1 to 6 if H is replaced by H^*. He will find that they all take a simpler form in view of the definition of $P(E)$.

T7 If E implies F and $P(E) \neq 0$, then $P(F \mid E) = 1$.
For $P(F \mid E).P(E) = P(E.F) = P(E)$ by A4.

Corollaries
(i) $P(H \mid H) = 1$ if $P(H) \neq 0$.
(ii) If H^* implies H then $P(H) = 1$, i.e. if H is certain then it is almost certain.
(iii) $P(H^*) = 1$. (This sharpens A5.)

T8 If $P(E) = 0$ then $P(E \mid H) = 0$ assuming that $P(H) \neq 0$.
For $P(E \mid H).P(H) = P(E.H) = P(E).P(H \mid E) = 0$, etc.

T9 If H implies $\bar{E}$ then $P(E \mid H) = 0$ if $P(H) \neq 0$. In particular if E is "impossible" then it is almost impossible. (The converse could hardly be true. This is intuitively clear in virtue of Section 2.8.)

† See Jeffreys, 1939, 29, and von Mises, 1942, for discussions of the terminology.
‡ In ordinary language a distinction is drawn between "hypotheses" and "theories"; hypotheses are improbable theories. This distinction is inconvenient for us and will be dropped. (See "Theory" in the index.)

PROOF. $E.H$ is a logically false proposition, and so by the definition of "implication" it follows that $E.H$ implies any proposition. In particular $E.H$ implies E^*. (See A6.) Now let us suppose that T9 is false, i.e. for some E and H, $P(E \mid H) \neq 0$. Then $P(E.H) = P(E \mid H).P(H) \neq 0$. Therefore by T7, $P(E^* \mid E.H) = 1$. But by A6 and T8, $P(E^* \mid E.H) = 0$, and this is a contradiction. So $P(E \mid H) = 0$.

COROLLARIES

(i) If $P(H) \neq 0$ then $P(E.\bar{E} \mid H) = 0$ (for $E.\bar{E}$ is logically impossible). In particular $P(E.\bar{E}) = 0$.

(ii) Let the phrase "E and F are mutually exclusive given H" mean (as in 2.3) "H implies the negation of $E.F$". Then if E and F are mutually exclusive given H, it follows that E and F are almost mutually exclusive given H, assuming that $P(H) \neq 0$.

(iii) Corollary (ii) may be extended in the obvious way to a finite set of propositions $E_1, E_2, E_3, \ldots$ Thus the word "almost" may be omitted in the statement of T2, if $P(H) \neq 0$.

T10 If $P(H) \neq 0$ then $P(E \vee \bar{E} \mid H) = 1$. In particular $P(E \vee \bar{E}) = 1$.

For H implies $E \vee \bar{E}$, whatever H may be, and the theorem follows from T7.

T11 If $P(H) \neq 0$ then $P(E \mid H) + P(\bar{E} \mid H) = 1$. In particular
$$P(E) + P(\bar{E}) = 1.$$

PROOF. By T10, $P(E \vee \bar{E} \mid H) = 1$ and by T9, cor. (i), E and $\bar{E}$ are almost mutually exclusive given H, so the theorem follows by the addition law A2.

COROLLARIES

(i) If $P(E \mid H) = 0$ then $P(\bar{E} \mid H) = 1$ and vice versa, assuming that $P(H) \neq 0$.

(ii) If F is independent of E given H and if $P(E.H) \neq 0$, then $\bar{F}$ is independent of E given H. (The condition $P(E.H) \neq 0$ implies $P(H) \neq 0$, by A3.)

T12 If $P(H) \neq 0$, then $0 \leqslant P(E \mid H) \leqslant 1$.

The first half of this inequality is simply A1. To prove the second half observe that by T11,
$$P(E \mid H) = 1 - P(\bar{E} \mid H) \leqslant 1,$$
by A1 again. (The assumption $P(E \mid H) \geqslant 0$ has not previously been used.)

T13 Suppose that E implies F. Then $P(F \mid H) \geqslant P(E \mid H)$, assuming that $P(H) \neq 0$.

PROOF. If $P(E \mid H) = 0$ there would be nothing to prove. On the other hand, if $P(E \mid H) \neq 0$ it may be shown, to begin with, that $P(F \mid H) \neq 0$. For suppose $P(F \mid H) = 0$. Then
$$\begin{aligned}P(E \mid H) &= P(E.F \mid H) \text{ by A4,} \\ &= P(F \mid H).P(E \mid F.H) \\ &= 0, \text{ by A1,}\end{aligned}$$

and this is a contradiction. Thus $P(F \mid H) \neq 0$. Therefore
$$P(F.H) = P(H).P(F \mid H) \text{ by A3},$$
$$\neq 0.$$
Therefore, by T12, $P(E \mid F.H) \leq 1$. But
$$P(E \mid H) = P(E.F \mid H) \text{ by A4},$$
$$= P(F \mid H).P(E \mid F.H).$$
Therefore $P(F \mid H) \geq P(E \mid H)$.

Definition. Any finite set of propositions E_1, E_2, E_3, ... such that $P(E_1 \vee E_2 \vee E_3 \vee \ldots \mid H) = 1$ is called *almost exhaustive* given H. If H implies $E_1 \vee E_2 \vee E_3 \vee \ldots$ then we say (as in **2.3**) that E_1, E_2, E_3, ... are *exhaustive* given H. In this case they are almost exhaustive given H if $P(H) \neq 0$, in virtue of T7.

T14   If the finite set of propositions E_1, E_2, E_3, ... are almost exhaustive given H and almost mutually exclusive given H, then
$$P(E_1 \mid H) + P(E_2 \mid H) + \ldots = 1.$$
This follows at once from T2.

T15   If E is equivalent to $E_1 \vee E_2 \vee \ldots \vee E_m$ where $E_1, E_2, \ldots, E_n$ are n mutually exclusive, equally probable and exhaustive propositions given H, where $P(H) \neq 0$, then $P(E \mid H) = m/n$. (This follows from T14 and T2.)

This theorem was to be expected in virtue of Section **2.3**. Observe that it does not *prove* the existence of probabilities other than 0 and 1. Thus the possibility is left open that every proposition can be proved or disproved by " pure thought ". (But see the second " suggestion " in **4.3**.)

T16   If $P(H) \neq 0$, then
$$P(E \vee F \mid H) + P(E.F \mid H) = P(E \mid H) + P(F \mid H).$$
PROOF.   Observe that $E \vee F$ is equivalent to $E \vee F.\bar{E}$,† so
$$P(E \vee F \mid H) + P(E.F \mid H) = P(E \vee F.\bar{E} \mid H) + P(E.F \mid H) \text{ by A4},$$
$$= P(E \mid H) + P(F.\bar{E} \mid H) + P(E.F \mid H) \text{ by A2 and T9},$$
$$= P(E \mid H) + P(F.\bar{E} \vee F.E \mid H) \text{ by A2, T9 and A4},$$
$$= P(E \mid H) + P(F \mid H) \text{ by A4}.$$

The above theorem is a generalisation of the addition law A2.

COROLLARIES

(i) If E and F are both almost certain given H, then $E.F$ is almost certain given H, if $P(H) \neq 0$. (This follows neatly from T12 and T16.)

---

† We are using the convention with regard to the omission of brackets which is analogous to that used in elementary algebra, a conjunction being the analogue of a product.

THE ABSTRACT THEORY

(ii) If $E_1, E_2, \ldots, E_n$ are almost certain given H, then so is their conjunction, if $P(H) \neq 0$. (By induction from cor. (i).)

(iii) If all the numbers $P(E_r \mid H)$ are either 0 or 1, then the formula of T5 holds. (Follows from cor. (ii) and T5A.)

(iv) $P(E_1 \vee E_2 \vee \ldots \vee E_n \mid H) \leqslant P(E_1 \mid H) + P(E_2 \mid H) + \ldots + P(E_n \mid H)$ if $P(H) \neq 0$. The case $n = 2$ is clear from T16 and the general result follows by induction.

(v) $P(E_1.E_2 \ldots E_n \mid H) \geqslant 1 - P(\bar{E}_1 \mid H) - P(\bar{E}_2 \mid H) \ldots - P(\bar{E}_n \mid H)$, if $P(H) \neq 0$.

For $P(E_1.E_2 \ldots \mid H) = 1 - P(\overline{E_1.E_2.\ldots} \mid H)$ by T11,
$= 1 - P(\bar{E}_1 \vee \bar{E}_2 \vee \ldots \mid H)$ by A4,
$\geqslant 1 - P(\bar{E}_1 \mid H) - P(\bar{E}_2 \mid H) - \ldots$ by cor. (iv).

T17 *The probability of a disjunction.* (Poincaré, 1912.) If $E_1, E_2, E_3, \ldots$ is any finite set of propositions and $P(H) \neq 0$ then
$P(E_1 \vee E_2 \vee E_3 \vee \ldots \mid H)$

$$= \sum_r P(E_r \mid H) - \sum_{r<s} P(E_r.E_s \mid H) + \sum_{r<s<t} P(E_r.E_s.E_t \mid H) - \ldots$$

This theorem is a further generalisation of the addition law, and it can be proved by mathematical induction from T16. It is often useful in difficult calculations.

T18 *The probability of a logical combination of propositions.* Let $E_1, E_2, E_3, \ldots, E_n$ be n propositions that are independent given H where $P(H) \neq 0$, and let $P(E_r \mid H) = p_r (r = 1, 2, \ldots, n)$. Let E be any combination of $E_1, E_2, E_3, \ldots, E_n$ by means of conjunctions, disjunctions and negations. Then $P(E \mid H)$ can be expressed as a function of $p_1, p_2, \ldots, p_n$.

PROOF. Let F_s ($s = 1, 2, \ldots, 2^n$) represent the various conjunctions similar to $E_1.\bar{E}_2.\bar{E}_3.\ldots E_n$, in which each term may or may not be negated. It is an elementary theorem † in symbolic logic that E can be expressed as a disjunction of some or all of the propositions F_s. Now the propositions F are mutually exclusive. Therefore $P(E \mid H)$ can be expressed as a sum of terms of the type $P(F_s \mid H)$, by T9, cor. (iii). Finally $P(F_s \mid H)$ can be expressed as a product; for example,

$P(E_1.\bar{E}_2.\bar{E}_3 \ldots E_n \mid H) = p_1(1-p_2)(1-p_3) \ldots p_n$.

If any of the factors $p_1, 1-p_2, 1-p_3, \ldots, p_n$ is zero, this is an immediate consequence of T5A. Otherwise it follows from T5 and T11. It is necessary to know that $E_1, \bar{E}_2, \bar{E}_3, \ldots, E_n$ are independent given H. This may be proved by an inductive argument, using T11 and its second corollary, together with the assumption that none of the factors is zero.

† See for example Hilbert and Ackermann, 1946, 16.

Example. To find $P(E \mid H)$ where $E = E_1 \vee (E_2 . \bar{E}_3)$. Here
$$E = \{(E_1.E_2.E_3) \vee (E_1.E_2.\bar{E}_3) \vee (E_1.\bar{E}_2.E_3) \vee (E_1.\bar{E}_2.\bar{E}_3)\} \vee \{(E_1.E_2.\bar{E}_3) \\ \vee (\bar{E}_1.E_2.\bar{E}_3)\}$$
$$= (E_1.E_2.E_3) \vee (E_1.E_2.\bar{E}_3) \vee (E_1.\bar{E}_2.E_3) \vee (E_1.\bar{E}_2.\bar{E}_3) \vee (\bar{E}_1.E_2.\bar{E}_3).$$
Therefore
$$P(E \mid H) = p_1 p_2 p_3 + p_1 p_2 (1-p_3) + p_1(1-p_2)p_3 + p_1(1-p_2)(1-p_3) \\ + (1-p_1)p_2(1-p_3)$$
$$= p_1 + (1-p_1)p_2(1-p_3).$$

The same result could be obtained by observing that E is equivalent to $E_1 \vee (\bar{E}_1 . E_2 . \bar{E}_3)$.

COROLLARY. The same methods may be applied even if $E_1, E_2, \ldots, E_n$ are not independent, provided that their probabilities (on the given evidence) are all 0 or 1.

To see this it is sufficient to use T16, cor. (iii), instead of T5.

This corollary may be used for the construction of " truth tables " in formal logic. Thus, in the previous example the formula $p_1 + (1-p_1)p_2(1-p_3)$, with p_1, p_2, p_3 all equal to 0 or 1, can be used to construct the truth table for the logical expression $E_1 \vee (E_2 . \bar{E}_3)$.

T19 Let $E_1, E_2, \ldots, E_n$ be independent given H, where $P(H) \neq 0$, and suppose that $P(E_1 \mid H) = P(E_2 \mid H) = P(E_3 \mid H) = \ldots = p$. Let F represent the proposition that exactly r of the E's are true, the other $(n-r)$ being false. Then
$$P(F \mid H) = \binom{n}{r} p^r (1-p)^{n-r},$$
where $\binom{n}{r}$ is the binomial coefficient $\dfrac{n!}{r!(n-r)!}.$

PROOF. The proof is essentially the same as in the last theorem. F can be expressed as the disjunction of $\binom{n}{r}$ propositions of the form
$$E_{m_1} . E_{m_2} . \ldots E_{m_r} . \bar{E}_{m_{r+1}} . \bar{E}_{m_{r+2}} . \ldots \bar{E}_{m_n},$$
where $m_1, m_2, \ldots m_n$ is some permutation of the suffixes $1, 2, \ldots, n$. These $\binom{n}{r}$ propositions are all mutually exclusive and the probability of each of them, given H, is $p^r(1-p)^{n-r}$. The result follows from T9, cor. (iii).

T20 Let the infinite sequence of propositions (" trials ") $E_1, E_2, \ldots$ be independent given H, where $P(H) \neq 0$, and suppose that $P(E_1 \mid H) = P(E_2 \mid H) = \ldots = p$. Let $F_{n,m,\varepsilon}$ be the proposition that
$$|f_n - p| < \varepsilon, \quad |f_{n+1} - p| < \varepsilon, \ldots, |f_m - p| < \varepsilon,$$
where f_n is the proportion of true propositions amongst $E_1, E_2, \ldots, E_n$ (with

similar definitions for f_{n+1} etc.). Then for any given positive numbers ε and τ, there exists n such that

$$P(F_{n,m,\varepsilon} \mid H) > 1 - \tau$$

for all $m \geqslant n$.

This theorem † corresponds to the frequency definition. An outline of the proof will be given. Observe that, for sufficiently large n,

$P(F_{n,m,\varepsilon} \mid H)$
$\geqslant P\{|f_n - p| < n^{-\frac{1}{4}}. \quad |f_{n+1} - p| < (n+1)^{-\frac{1}{4}}. \ldots |f_m - p| < m^{-\frac{1}{4}} \mid H\}$
$\geqslant 1 - \sum_{\nu=n}^{m} P(|f_\nu - p| \geqslant \nu^{-\frac{1}{4}} \mid H)$ by T16 cor. (v).

It can be shown, by using T19 together with some analysis,‡ that

$$P(|f_\nu - p| \geqslant \nu^{-\frac{1}{4}} \mid H) < K\nu^{-2},$$

where K depends only on p. The theorem now follows at once.

If the axiom of complete additivity is assumed this theorem can be shown to be "equivalent" to a theorem due essentially to Borel,§ that *it is almost certain that the proportion of " successes " in the first n " trials " tends to p as $n \to \infty$*. Since an infinite number of trials cannot be completed in practice there is much to be said for T20 in spite of the complicated wording. This exemplifies a point made above in connexion with the axiom of complete additivity, namely that it is mathematically convenient but is not essential for the applications.

A similar result to T20 could of course be proved corresponding to von Mises' assumption concerning subsequences. (See 1.4, ii.)

Summary. A fairly detailed theory has been deduced from six purely formal axioms. Within this abstract theory there are results corresponding (verbally) to the two classical definitions of probability. The correctness of the theorems does not depend on any philosophical interpretation of probability.

† There is a very similar theorem due to Cantelli. See Uspensky, 1937, 101. A result usually known as "Bernoulli's theorem" is the special case of T20 with $m = n$.
‡ Cf. M. Fréchet, 1937, 217-22. The analysis is not trivial. It depends on the approximation of $\sum_{r=\varepsilon}^{\nu} \binom{\nu}{r} p^r (1-p)^{\nu-r}$ by means of an error function. (See 5.3.)

Chapter 5 of Fréchet's book contains an account of generalisations of T20 due to F. P. Cantelli, A. Kolmogoroff, A. Khintchine and Paul Lévy. See also W. Feller, 1945.
§ See Fréchet, 1937, 216 and 228-31. Any two mathematical theorems are "equivalent" in the sense of A4. Here we mean that the number of mathematical steps required is not large.

3.4 An alternative set of axioms

Consider the axioms:

B1 $P(E)$ is a non-negative number,
B2 $P(E \vee F) = P(E) + P(F)$ if $P(E.F) = 0$,
B3 if E implies F then $P(F) \geqslant P(E)$,
B4 $P(H^*) \neq 0$,
B5 $P(E^*) = 0$ for some proposition E^*,

together with the definition

$$P(E \mid H) = P(E.H)/P(H) \quad \text{if } P(H) \neq 0.$$

These are all consequences of the previous abstract theory, and it is easy to see, conversely, that they imply axioms A1 to A6 if " almost impossible " propositions are not allowed to occur to the right of the vertical stroke.

The self-consistency of the axioms B1 to B5 is seen at once by imagining all propositions to be true or false and calling their probabilities 1 or 0 respectively. This does not prove the self-consistency of the system of axioms obtained by adding an axiom to the effect that there is at least one proposition whose probability is not 0 or 1.

The new set of axioms is more economical than the old set. But Chapters 1 and 2 do not directly suggest the new axioms. The symbols $P(E)$ etc. correspond to those beliefs that are most liable to be regarded as meaningless,† and the probabilities that are easier to interpret as reasonable beliefs are introduced merely by way of a definition. It is for this reason that we preferred to start from axioms A1 to A6. Of course these axioms also involve numerical values for symbols like $P(E)$ where E is empirical. It may therefore be felt that they achieve too much, for they attach a meaning to a probability that may not correspond to a reasonable belief. But this does no harm; in fact it is actually an advantage since the use of symbols like $P(E)$ simplifies the calculations in some problems. (The reader should refer back to the modified definition of probability given in a footnote to **1.3**. See also the remarks about " unobservables " in **4.4**.)

† Cf. **1.2**.

CHAPTER 4

THE THEORY AND TECHNIQUE OF PROBABILITY

> " It is no paradox to say that in our most theoretical moods
> we may be nearest to our most practical applications."
>
> A. N. WHITEHEAD

THE abstract theory of the previous chapter is a branch of pure mathematics in which it is unnecessary to attach any non-mathematical meaning to the word " probability ". Once an abstract theory has been developed there arises the highly controversial question of how the theory is to be applied. This question forms the subject-matter of the present chapter. It will be necessary to restore the meaning of " probability " that was given in **1.3**.

It will be convenient to distinguish between " axioms ", " rules " and " suggestions ". The axioms are the assumptions of the abstract theory. The " rules " connect this abstract theory with actual or hypothetical judgments concerning degrees of belief. These rules are listed in **4.1**. The deductions from the combined axioms and rules constitute the " theory of probability ". Finally the " suggestions " are natural modes of procedure for forming bodies of beliefs. Some of them are given in **4.3**. There is no compulsion to accept them in order to be able to use the theory. The consequences of accepting the axioms, rules and suggestions may be called the " technique of probability ". This technique will not be completely defined since no complete list of suggestions will be given.

The suggestions emerge from a familiarity with the theory and applications of probability. For example, *any general theorem of the abstract theory may influence what you regard as correct to assert as your own $\mathcal{B}$*. It is therefore impracticable to list all possible suggestions.

A drawback of some existing theories is that they are not " theories " in the above sense; i.e. the axioms, rules and suggestions are not distinguished. This makes it difficult to separate any large part as belonging entirely to the realm of logic and mathematics.

The trichotomy into axioms, rules and suggestions is perhaps the ideal form for any scientific theory.

4.1 The " rules "

(i) An expression of the form $P(E \mid H)$ is given a double interpretation. First it is regarded as a number subject to the axioms of the abstract theory, and second as a reasonable belief in E when H is assumed, if this belief has any meaning. There is no necessity to insist that H should be known to be true; in fact the applications would thereby be much restricted.

(ii) Relations like $P(E \mid H) > P(E' \mid H')$, $P(E \mid H) < P(E' \mid H')$, $P(E \mid H) = P(E' \mid H')$ also have two interpretations. They may be regarded as ordinary arithmetical relations, or else as assertions that one reasonable belief is (for example) more intense than another, provided that you consider that both sides of the comparison have a meaning. (Cf. 1.2.) The possibility is not ruled out that the theory will throw up some meaningless comparisons.

(iii) A body $\mathcal{B}$ of beliefs consists of a set of inequalities and equalities between probabilities. Some or all of these may be written down by a person's direct intuitive judgment, or they may simply be assumed. Some of the judgments may be "laws of nature". Generalisations of this definition of $\mathcal{B}$ will be discussed in 4.12.

(iv). Deductions may be drawn by using the abstract theory together with $\mathcal{B}$. Those deductions that are of the form of inequalities or equalities between probabilities may have an intuitive significance.

(v) If a contradiction is reached, $\mathcal{B}$ is said to be inconsistent or unreasonable.

(vi) Rule (iv) may give rise to intuitive relations that are not already included in $\mathcal{B}$. *These may be added to $\mathcal{B}$, thereby forming a larger body of beliefs which may also be denoted by $\mathcal{B}$.*

(vii) Logically it would be better if we used two different symbols, say $P(E \mid H)$ and $P'(E \mid H)$, for the two different meanings. Then rule (ii) could be expressed by saying that the inequality

$$P(E \mid H) > P(E' \mid H')$$

implies and is implied by the comparison

$$P'(E \mid H) > P'(E' \mid H')$$

and so on. (The second sign " $>$ " means " is more intense than ".) But a gain in logical rigour is not always a gain in clarity. Hence only one notation will be used instead of two. This will enable the arguments to be expressed more briefly.

(viii) If $\mathcal{B}$ contains no judgments, none can be deduced. Thus the theory cannot be applied without some intuitive interpretation of probability.† This is again analogous to the applications of geometry or of any other abstract science.

(ix) Notice that the theory can be applied to any body of beliefs, but the application is of practical importance only if the body of beliefs is accepted by some individual.

† This shows that our theory of probability is not an objective one in the sense of Section 1.3 (i.e. "constructibly objective").

4.1A The justification of the theory

The exposition of the foundations of the proposed "theory" has now been completed. It should be very carefully noticed that there is no claim that reasonable beliefs can be measured in general—only that relations can be stated between them. In fact, it seems to the writer that the theory involves about as many relations as it is possible to state in a precise manner. No doubt the theory can be supplemented by means of suggestions, but these are not precise (and they belong to the "technique" rather than to the "theory").

The question arises to what extent the theory can be justified *a priori*, that is, before making practical use of it. To this end the following exceedingly crude argument is proposed.

Suppose first that it is always possible to apply to $P(E \mid H)$ the definition by equally probable cases, at least as an arbitrarily good approximation,† and assuming that H is not impossible. It would be surprising if it were possible to *prove* that this cannot be done. An inconsistency within the abstract theory would amount to such a proof. Therefore the abstract theory is presumably consistent, even with the assumption that probabilities other than 0 or 1 occur. (Cf. **3.4**.)

Now it is natural, I think, to assume that any reasonable‡ $\mathcal{B}$ would be *consistent* with the possibility that the definition by equally probable cases was applicable, even though $\mathcal{B}$ may not be dependent upon this definition. Then such a $\mathcal{B}$ cannot lead to a contradiction when combined with the theory; in other words $\mathcal{B}$ must be "reasonable" in the technical sense. The fact that no contradiction is obtained may not be regarded as sufficient justification for accepting the theory. But suppose that when the theory is combined with a reasonable $\mathcal{B}$ it leads to a "comparison" of the form $P(E \mid H) > P(E' \mid H')$. Then, since no contradiction can be obtained, we know that, in an enlarged § $\mathcal{B}$, $P(E \mid H) > P(E' \mid H')$ *if $P(E \mid H)$ and $P(E' \mid H')$ can be compared*. It seems natural from this to assert simply $P(E \mid H) > P(E' \mid H')$ when this comparison means anything. (In order to be convinced of this last step the reader should consider an example.) This is equivalent to accepting the theory.

4.2 Inaccurate language

In most applications of probability the propositions E and H in the expression $P(E \mid H)$ are in a form describing a physical situation. Accordingly we shall often talk about the *probability of an event* when we mean the probability

---

† We are here implicitly taking a result like T13 for granted, and the "approximation" is supposed to be of the form that a probability lies in a narrow interval (with rational end-points).
‡ The word "reasonable" is used here in a non-technical sense for once.
§ See rule (vi) in **4.1**.

4.3 Some "suggestions"

of a proposition asserting that the event will happen or has happened. Various other rather inaccurate forms of language will be used without explanation. This is necessary in order to save space and to avoid cumbersome phrases.

4.3 Some "suggestions"

The classification of the fundamentals of probability into axioms, rules and suggestions has already been discussed. The mathematical theory depends only on the axioms. The rules are not purely mathematical, but they are precisely stated in terms of the primitive notion of the comparison of pairs of beliefs. They enable the mathematical theory to be applied to a given body of beliefs. The " suggestions " are liable to affect your body of beliefs without directly using the theory, and the present section contains some examples of this. It does not seem to be possible to formulate the suggestions with the same precision as the axioms and rules. Non-mathematical words such as " honesty " will be used.

The rejection of any of the suggestions would have no effect on what we have called the " theory of probability ".

(i) *Numerical probabilities.* It will be recalled that the axioms were largely derived by imagining perfect packs of cards. Having accepted the axioms themselves it is natural to accept the notion of a perfect pack of cards. This provides a significance for all numerical probabilities that are rational numbers between 0 and 1 (and therefore also for the irrational numbers). If real packs of cards are preferred they serve the same purpose, but the probabilities are then best regarded as in some sense good approximations. (See **4.6**.)

If it is taken for granted that $\mathcal{B}$ contains all the obvious judgments concerning packs of cards, then it becomes intelligible to accept as a probability judgment any numerical statement such as $\frac{1}{2} < P(E \mid H) < \frac{3}{4}$ or $P(E \mid H) = \frac{1}{2}$. Moreover, with practice it may be possible to make such judgments without thinking of a concrete example of probabilities of $\frac{1}{2}$ and $\frac{3}{4}$. There is an analogy with the judgment of distances. A very young child can judge that one line is longer than another one before he can associate a distance with a number of inches.

It is not obvious whether it is ever reasonable to judge that a probability is precisely equal to a definite number such as $\frac{1}{2}$. But it may often be judged that such an equality is a sufficiently good approximation for some particular purpose. In such cases we shall say that the probability is $\frac{1}{2}$, without troubling to add that the judgment is intended only as an approximation.

(ii) *Empirical propositions.* A particular case of numerical probabilities is given by probabilities of 0 and 1. Now if E is an empirical proposition rather than a logical one, is it possible to have $P(E \mid H) = 0$ or 1 exactly ? The answer to this question is suggested by T8 and T11, cor. (i). These results show that if $P(E \mid H) = 0$ or 1 then no amount of additional evidence can change the probability of E unless the additional evidence is itself almost impossible, given H.

The suggestion that emerges from this is that an empirical proposition cannot be almost certain (in the technical sense of course) unless it is logically implied by the evidence. If E is logically implied by H then it is certain, assuming H—not merely almost certain. Almost certainty that is not actual certainty seems to occur only in purely mathematical examples. These may, however, be convenient models of practical problems.

The suggestion that the probabilities of empirical propositions cannot have the values 0 or 1 is taken as an axiom by Jeffreys. This course has not been followed here since the abstract theory can be built up satisfactorily from the axioms given in Chapter 3.

(iii) *The device of imaginary results.* The idea behind the previous suggestion can be extended into a very useful technique for helping you to arrive at inequalities for probabilities in difficult cases.

Suppose, for example, that you wish to estimate the initial probability † that a man is capable of extra-sensory perception, in the form of telepathy. You may imagine an experiment performed in which the man guesses 20 digits (between 0 and 9) correctly. If you feel that this would cause the probability that the man has telepathic powers to become greater than $\frac{1}{2}$, then the initial probability must be assumed to be greater than 10^{-20}. (This follows by a simple application of Bayes' theorem: cf. **6.1**.) Similarly, if three consecutive correct guesses would leave the probability below $\frac{1}{2}$, then the initial probability must be less than 10^{-3}.

(iv) *Honesty.* A suggestion which seems obvious enough is that in order to avoid ultimate contradictions all probability judgments should be honestly held, and should be arrived at unemotionally.

There is an apparent exception to this suggestion. You may sometimes work with a simplified form of $\mathcal{B}$. When this is done there should be a judgment that it will lead to sufficiently good results for the purpose in hand. This is an example of the usual scientific method of "idealising" a problem. There is no real dishonesty in the procedure, provided that it is not claimed at the end of the calculations that the results follow from the original unsimplified $\mathcal{B}$.

(v) *The classical definitions.* Theorems T15 and T20 make both the classical definitions ‡ of probability relevant as a guide to probability judgments. (See also paragraph (d) on page 12 and Sections **4.10** and **4.11**.)

(vi) *The design of experiments.* The interpretation of the results of an experiment always depends on the judging of probabilities. It is sometimes

---

† i.e. the probability before some experiment is performed.
‡ Namely the frequency definition and the definition by equally probable cases.

possible to design an experiment so that the intervals in which the probabilities are judged to lie are narrow rather than wide. Other things being equal, such a design is to be recommended. For applications of this suggestion the reader is referred to R. A. Fisher's *The design of experiments* (5th edn., 1949).

4.4 A non-numerical theory

The assumption that $P(E \mid H)$ is a number † is largely for mathematical convenience. There may be no way of deciding at all precisely what this number is. This method of assuming the mathematical existence of " unobservables " is familiar in modern physics and in philosophy. (It was pointed out in **1.4** (viii) that a theory can be constructed without the assumption that probabilities can be represented by numbers.) The assumption of the " existence " of an unobservable means that all observable and all meaningful deductions must be accepted. (Cf. **3.4**.)

4.5 Practical difficulties

Difficulties arise in all applications of mathematics (and elsewhere) because practical problems are usually very complicated. In the theory of probability it often happens that you are interested in $P(E \mid K)$ where K represents everything you know. In this case it is out of the question to list K as a collection of precise statements, especially as your knowledge contains much that is half-forgotten. Similarly it may be inconvenient to define E very precisely. For example, if you are interested in the probability of rain, you do not usually specify how much water must fall before it is called rain. On the other hand, all those judgments in ℬ that are used in the course of any discussion can be clearly stated in terms of the propositions E, K etc., even though these propositions are themselves not completely defined.

Usually most of K is judged to be more or less irrelevant. It may be possible to state the relevant part, H, with a fair degree of precision. You may then prefer to work with $P(E \mid H)$ and to regard it as roughly ‡ the same as $P(E \mid K)$. (It is precisely this process which is used in law courts when " hearsay evidence " is ignored.) It is worth emphasising that such complications and approximations are inevitable in applied mathematics. Any discussion which does not recognise them is simply incomplete. (See also **4.3** (iv).)

4.6 The principles of " insufficient reason " and " cogent reason "

Let G be the proposition " I have just spun a coin and allowed it to fall to the ground." Let H be the proposition that " heads " is uppermost. Can the

---

† The symbol " P " here has the meaning of " P " rather than of " P' ". See rule (vii) of **4.1**.

‡ This approximate equality between $P(E \mid H)$ and $P(E \mid K)$ is a probability judgment belonging to ℬ.

reader state a relation of equality or inequality between his degrees of belief $P(H \mid G)$ and $P(\bar{H} \mid G)$? In accordance with the preceding section no precise description will be given of how the coin was spun, but it may be assumed that there is no "catch". The following replies (amongst others) may be given by different readers.

(i) $P(H \mid G) > P(\bar{H} \mid G)$ by "extra-sensory perception".

(ii) No opinion offered.

(iii) $P(H \mid G) = P(\bar{H} \mid G)$ because there is absolutely no reason to expect one of H or $\bar{H}$ rather than the other. This is an application of the "principle of insufficient reason", also known as the "principle of indifference".

(iv) $P(H \mid G) = P(\bar{H} \mid G)$ because the problem is physically symmetrical with respect to heads or tails. This is an application of the "principle of cogent reason".†

(v) $P(H \mid G)$ is approximately equal to $P(\bar{H} \mid G)$, the approximation being very close because the problem is very nearly symmetrical.

(vi) More precisely the difference between $P(H \mid G)$ and $P(\bar{H} \mid G)$ is less than $1/1000$.

Observe that (v) and (vi) make direct use of the *numerical* concept of probability. But it is possible to modify them a little, so as to avoid this, by introducing a subsidiary event E which is very improbable on the evidence G. E might be that I had lost the coin while spinning it and it could be judged that (a) $P(\bar{H}.\bar{E} \mid G) < P(H \mid G)$, and (b) $P(E \mid G)$ is less than the probability of selecting a specified card from a pack containing 1000 cards.

But in future such tedious interpretations will be avoided. Instead a bold use will be made of numerical probabilities, both in the statement of ℬ and in the answers to problems. It is emphasised once for all that these numerical probabilities can be given at least a partial interpretation in terms of inequalities between pure degrees of belief. Life is too short to give these interpretations on every occasion. One simple way of supplying the interpretations when required is by using packs of cards as in 4.3.

As regards the alternative judgments (i) to (vi), the theory gives no way of deciding between them as they stand. My own preference is for alternative (vi). Number (iv) may be more appropriate for the idealised problem in which the real coin is replaced by a perfect one. And even for the real problem it is more convenient to *assert* number (iv) and *mean* number (v) or (vi). Such a policy will sometimes be adopted in future.

† Russell (*Human knowledge*, 397) formalises the principle thus:
$$P\{\phi(a) \mid \psi(a)\} = P\{\phi(b) \mid \psi(b)\},$$
where ϕ and ψ are propositional functions not involving a or b. In the present theory the principle hardly requires formalising because if the formalism were judged to be (approximately) applicable, the probabilities would be judged to be (approximately) equal without reference to the formalism.

4.7   PROBABILITY AND WEIGHING OF EVIDENCE

4.7 Simple examples

(i) n people are chosen "*at random*". What is the probability that no pair of them will have the same birthday? Assume for simplicity that there are 365 days in the year.

First we must say what is meant by selecting n people "at random". It means that out of some population, say the population of England at a given time, each person in the population has an equal probability of being selected. One method of making such a selection is to construct a "model" of the population consisting of cards, one card for each person in the population. A selection of n cards may be made by a process that is judged to be random.† The people are then taken corresponding to the cards selected. The process of taking n things at random out of a "population" is called "taking a sample" or more precisely "*taking a random sample*". In our example the sample is one "without replacement" since it is specified that the n people are all different.

Let us suppose that you *know* the number of people born on each day of the year in the entire population, and let the proportions of those born on the 1st, 2nd, 3rd ... days of the year be $p_1, p_2, p_3, \ldots, p_{365}$. By T15 these are the probabilities of the first person selected being born on the 1st, 2nd, 3rd ... days of the year. If the population is large the probabilities for the second person will be effectively the same even if you are told the first person's birthday, and so on for all n people. Hence by T5, the probability that the birthdays of the 1st, 2nd, ... persons are respectively on the r_1th, r_2th, ... days is $p_{r_1} \cdot p_{r_2} \ldots p_{r_n}$. Therefore the required probability is the sum of all such expressions with unequal suffixes.‡ This uses T1 or T9, cor. (iii), depending on whether a definition is supplied for the birthday of a person born exactly at midnight. (This type of hair-splitting will be ignored in future.)

It is not difficult to prove the (intuitively reasonable) fact that the probability will be a maximum when $p_1 = p_2 = \ldots = 1/365$. Thus the required probability is less than or equal to $n! \binom{365}{n} 365^{-n}$. With $n = 23$ the probability is less than $\frac{1}{2}$. (The special case $p_1 = p_2 = \ldots = p_{365}$ is mentioned by H. S. M. Coxeter, *Mathematical recreations and essays*, 11th edn., 1940, London, p. 45. He attributes the result to H. Davenport, who, however, disclaims originality.)

(ii) Imperfect dice A and B are thrown twice and give scores a, a' and b, b', but these scores are not disclosed. Suppose that the probabilities of the various scores are $p_1, p_2, \ldots, p_6$ for die A and $q_1, q_2, \ldots, q_6$ for die B,

† Complete randomness may be unobtainable.
‡ In other words, it is $n!$ times the elementary symmetric function of the nth degree formed from the numbers $p_1, p_2, \ldots, p_{365}$.

and let the natural assumptions about independence be made. Then it is likelier that $a = a'$ and $b = b'$ than that $a = b$ and $a' = b'$. (This is reasonable intuitively, by a rough argument not involving a calculation. The result follows from the Cauchy-Schwartz inequality $\Sigma p_r^2 \Sigma q_r^2 \geqslant [\Sigma p_r q_r]^2$.)

Observe that here the probabilities p_1, q_1 etc. are given as part of the assumed body of beliefs. Therefore, as far as we have gone, there is no need to show how these probabilities could have been estimated. The result does not depend on the values of p_1, q_1, ... but only on their existence. Hence the result follows from a body of beliefs containing only the independence assumptions, just as in example (i).

4.8 Certainty and the " verification " of the theory

If a number of samples of ordinary air are taken, the proportions of oxygen in them will not all be exactly the same, though the differences may be too small to measure. There is an extremely small probability † that a large sample of air would contain no oxygen at all. It is theoretically possible that a man could die of suffocation as a consequence of this. Or that a particular man should win the Irish sweepstake every year for fifty successive years. In these cases it would be natural to say that a miracle had happened, or that there had been foul play. Under normal assumptions it would be virtually certain that they would not happen. Thus in addition to (logical) certainty and " almost certainty " there is such a thing as practical certainty. There are many other shades of meaning that are attached to the word " certain " in ordinary conversation.

The idea of practical certainty can be used in order to verify the theory of probability, or rather in order to show that it works. (To demand more than this would be like demanding a proof of a logical system.) A particular level of probability, very close to one, is selected somewhat arbitrarily, say $1 - 10^{-20}$. Then if $P(E \mid H) > 1 - 10^{-20}$ and if you know that H is true, you say ‡ that E will not be found to be false. In other words you make a definite prediction about E. If E is later found to be true you may say that the theory of probability has had some verification. If E is found to be false *you look to see if ℬ can be modified*, since it may have been written down carelessly in the first place.

There is a small point connected with the idea of certainty that will now be considered. Suppose that E is logically certain given H, i.e. that H implies E. Then we know by T7 that $P(E \mid H) = 1$, provided that H is not almost impossible. It could be assumed as a convention that $P(E \mid H) = 1$, even if H

† According to most theories of probability. Some people would assert that such small probabilities are meaningless. On this view some small number must exist below which probabilities may be regarded as zero. *A similar view has been propounded for numbers themselves.* The view would lead to unpleasant complications.

‡ At any rate most people would.

is almost impossible † (and similarly that $P(\bar{E} \mid H) = 0$). We know that this would lead to contradictions if H were allowed to be strictly impossible (see **3.1** (iv)). But if H were almost impossible though not strictly impossible, the convention would probably not lead to trouble. It would give us a little more freedom in purely mathematical problems connected with "geometrical probabilities".

In future it will be assumed, unless otherwise stated, that the "given" proposition H is not almost impossible, in expressions of the form $P(E \mid H)$.

4.9 Deciding between alternative hypotheses or scientific theories

If it is desired to decide which of two or more alternative hypotheses is likely to be correct in the light of experimental results, then the natural method is to use Bayes' theorem, T6. Objections have frequently been raised against Bayes' theorem on the grounds that the initial probabilities of the hypotheses cannot be estimated, or that they do not exist. The view held here is that the initial probabilities may always be assumed to exist within the abstract theory, but in some cases you may be able to judge only that they lie in rather wide intervals. This does not prevent the application of Bayes' theorem: it merely makes it less effective than if the intervals are narrow.

It is hardly satisfactory to say that the probabilities do not exist when the intervals are wide, while admitting that they do exist when the intervals are narrow.‡ This is, however, quite a common practice even when the interpretation is in terms of degrees of belief. There may be some convenience in the practice, but it is out of place in a discussion of fundamentals, and it will not be adopted here.

If, after the evidence is taken into account, it is found that a hypothesis H_1 is more probable than another one, H_2, this by itself will not necessarily make H_1 preferable to H_2. It is important also to allow for the utilities of H_1 and H_2, at least in some circumstances. For suppose that H_2 is an elaboration of H_1 so that it certainly implies H_1. Then the final probability of H_1 exceeds that of H_2 (though possibly by only a little), but H_2 may be much more useful and interesting. (This is particularly clear if H_1 happens to be H^*.) If, on the other hand, H_1 and H_2 are mutually exclusive, their utilities will not usually enter so decisively into consideration.

The alternative hypotheses may be scientific theories, one of which is assumed to be right.§ Bayes' theorem is therefore available as a method for

---

† If H is almost impossible we have not even proved that $P(E \mid H) \leqslant 1$.

‡ It would be forgivable to define the "meaningfulness" of a probability by means of the narrowness of the interval.

§ Often when it is said that a theory is "right" it is meant that it is in some sense a good approximation, and for the application of Bayes' theorem the sense must be defined. This must be done in such a way that the theory has no exceptions, otherwise its final probability will be zero. Remarks having some bearing on the initial probability of a theory will be found in **5.4** and **7.5**.

making advances in theoretical science. (It is the method of scientific induction in a numerical form.) But the question arises : what if the theories themselves involve probability statements (and they very often do) ? According to the convention at the beginning of **3.1** such theories cannot be considered as propositions. Let us call them " improper theories ", those that are expressible as propositions being called " proper theories ". (Similarly we can talk about proper and improper hypotheses and propositions.) It is not immediately clear how the theory of probability can be used for deciding between improper theories. Perhaps the most obvious method would be to extend the meaning of the word " proposition " so as to allow it to refer to probabilities, but this course may lead to logical difficulties.† (See **3.1** (vi).)

Sometimes the difficulty can be avoided by converting an improper theory into a proper one. For example, in the Mendelian theory of heredity, probabilities may be stated for an individual to have various characteristics, given those of its ancestors. In this form the theory is an improper one and it might contain a probability statement of the form $P(E \mid H) = p$. But let U be the proposition that animals or plants have chromosomes and genes. The chromosomes are assumed to occur in symmetrical pairs, and this symmetry leads to the judgment that $P(E \mid H.U) = p$. This judgment can be regarded as belonging to the body of beliefs, rather than to the theory of heredity. Thus the theory can be converted into a proper theory, namely the proposition U. This is really an over-simplification. It is possible that it would be judged that there might be a bias against the survival of one rather than the other form of a gene. The technique for dealing with this complication would be of the same kind as the one exemplified below in connexion with " extrasensory perception ". If it is assumed that there is no " bias " then the probabilities that occur are independent of any further experiments. Such probabilities are described by the technical term *chances*. The meaning of the term is made clearer by considering an unbiased coin to be spun a number of times. The fact that the coin is described as unbiased means that you have judged that its probability of coming down heads is $\frac{1}{2}$, and that this probability is a chance in the sense that it is independent of how many heads and tails have already been obtained.

The probabilities that occur in scientific theories are usually chances. Another example is afforded by quantum theory, in which the probability of a

† It may require a " theory of types ", as in symbolic logic.
Another way in which the difficulty arises is if you are interested in $P(E \mid H)$ where H consists of all known information, so that H must include the fact that you are interested in $P(E \mid H)$. This point will be ignored in the present book. It is important, however, when E depends on your own volition or imagination. Consider, for example, the probability that you will smoke within the next half-hour (given all known information). A similar point arises in politics, when a public forecast of an event may affect the probability of the event.

particle appearing in a volume of space is given by the integral over that volume of the square of the modulus of the appropriate wave function.† Here there is no method known of converting the theory into a proper theory. If it is ever possible to do this it would mean that quantum theory could be stated as a proposition U, where U asserts that the real universe is the same as some hypothetical universe $\mathfrak{U}$, whose relevant properties could be described without reference to probability. Any probability statement in quantum theory, of the form $P(E \mid H) = p$ could then be replaced by $P(E \mid H.U) = p$, and it could be transferred to $\mathfrak{B}$. The problem of the truth or falsehood of quantum theory would be replaced by that of U. Provisionally U may be regarded as the proposition " quantum theory is true ". ‡

In general, any improper theory can be formally converted into a proper theory in this way, by introducing a symbol U which is incompletely defined. This artifice is not very satisfactory, but it seems to be adequate for the applications.

It is often convenient to talk as if U were an objective description of some aspect of the physical world, without actually completing the definition of U and thereby expressing it as a proper theory. The only essential property of U is that $P(E \mid U.H)$ has known values for some propositions or " experiments " E, these values being the same for all reasonable bodies of belief. A number of theories of probability have been proposed in which such objective probabilities are the only admissible ones. Such theories are used by many leading statisticians. (See heading ix of 1.4.) From our point of view these theories are incomplete. They are essentially included in the present theory by the device of using incompletely defined propositions.

An objective probability, in the present theory, may also be described as " tautological ", i.e. its numerical value is known (usually precisely) because of the conventional manner of using incompletely defined propositions.§ When a tautological probability $P(E \mid U.H)$ is also a chance, then for all reasonable bodies of belief, the proportion of successes will almost certainly tend to $P(E \mid U.H)$ in an infinite sequence of trials, provided that U and H are true. Hence such a probability may be described as a " statistical probability ", and is so described for example by Bartlett (1936 and 1940).

---

† This has been denied by Jeffreys, 1942.

‡ It is only in virtue of the above formal artifice that it is legitimate to regard " quantum theory is true " as a proposition. The artifice can be avoided by the adoption of the generalised meaning of a proposition, discussed in 3.1 (vi).

§ A probability which is deduced by means of the abstract theory from tautological probabilities alone may also be called a " tautological probability ". A probability may of course be only partly tautological. Such a probability cannot occur in a dualistic theory in which tautological and non-tautological probabilities are given different notations, unless a third notation is introduced.

A chance can be cross-classified in two ways : (i) the " given " propositions may be true or false, (ii) the chance may be tautological or non-tautological. Thus there are four kinds of chances. It is usual to use the word " chance " for a *true* chance. A statistical probability is a tautological chance, not necessarily a true one.

The above discussion is in no way restricted to scientific theories in the ordinary sense. Suppose, for example, that you know that there are N adult males in England, and let U_M denote the proposition that M of them are over six feet high. Let E be the proposition that the next man selected will be over six feet. Suppose that the men are selected at random (see **4.7**). Then $P(E \mid H.U_M) = M/N$, where H is a description of the method of selection. There are $N+1$ possible theories concerning the value of M. A typical one of these could be stated as an *improper* theory in the form "$P(E \mid H) = M/N$". The proper theory corresponding to this is of course U_M. Notice that $P(E \mid H.U_M)$ is a chance if the sampling is with replacement. The equation $P(E \mid H) = M/N$ is generally false even if U_M is true. This suggests that in the general case it is quite essential to introduce U. For the probability statements of the improper theory are liable to contradict judgments already in your body of beliefs. If U_M is true, $P(E \mid H.U_M)$ may be called " the true probability of E given H ", but this mode of expression is misleading and is best avoided. It may, however, be called " the (true) chance " without serious risk of confusion.

It is sometimes convenient to make assertions like " the probability is $\frac{1}{2}$ that the chance of success is $\frac{1}{4}$ ". This assertion can be given a meaning in the same way that an improper theory can be converted into a proper one. It means " the probability is $\frac{1}{2}$ that H is true, where the chance of success, given H, is $\frac{1}{4}$ according to $\mathcal{B}$ ". In fact the rest of the discussion of the present section is really an attempt to attach a significance to the *probability of a chance*.

Let us consider in detail an example of the problem of deciding between " alternative bodies of belief ". This is of course the same in principle as deciding between improper theories.

Suppose that a coin is spun 1000 times and that the results are successively guessed. Let E_n mean that the guess of the nth spin is correct. Let $\mathcal{B}_1$ consist of the following judgments :—

(a) $P(E_n \mid H) = \frac{1}{2}$, where H is a description of how the experiment is performed;

(b) $E_1, E_2, \ldots$ are independent given H.

Let $\mathcal{B}_2$ be the same as $\mathcal{B}_1$ except that $P(E_n \mid H) = \frac{1}{2}$ is replaced by $P(E_n \mid H) = \frac{3}{4}$. Suppose that the number of successes is 497 out of 1000. Call this result E. In virtue of T20 (with $m = n$) you may be tempted to say that $\mathcal{B}_1$ is better than $\mathcal{B}_2$ or even that $\mathcal{B}_1$ is more probable than $\mathcal{B}_2$. These statements are illegitimate

4.9   PROBABILITY AND WEIGHING OF EVIDENCE

since $\mathcal{B}_1$ and $\mathcal{B}_2$ are not propositions. But now let us introduce a new proposition, K, which means that the man who is guessing has "extra-sensory perception" † (assumed permanently operating), and for $\mathcal{B}$ take the judgments :—

   (a) $P(E_n \mid H.\bar{K}) = \frac{1}{2}$ where H is a description of the experiment and includes a description of the man,

   (b) $P(E_n \mid H.K) = \frac{3}{4}$,

   (c) $E_1, E_2, \ldots$ are independent given $H.\bar{K}$, i.e. the probabilities in (a) are chances,

   (d) $E_1, E_2, \ldots$ are independent given $H.K$, i.e. the probabilities in (b) are chances,

   (e) $10^{-20} < P(K \mid H) < 10^{-3}$. (See **4.3** (iii).)

From these judgments and from the abstract theory it is quite easy to calculate $P(K \mid E.H)$, the new probability of the man having extra-sensory perception in virtue of the experiment E. The calculation (based on more natural assumptions) will be given in **6.5** and **7.3**. The result may be regarded as the answer to the original question of whether $\mathcal{B}_1$ is better than $\mathcal{B}_2$.

The assumptions are made more natural if it is supposed that K is the disjunction of a large number, k, of different propositions $K_1, K_2, \ldots, K_k$ where

   (a) $P(E_n \mid H.K_\kappa) = \frac{1}{2}(1 + \kappa/k)$,

   (b) $10^{-20}/k < P(K_\kappa \mid H) < 10^{-3}/k$,

   (c) $E_1, E_2, \ldots$ are independent given $H.K_\kappa$ for each κ. Instead of using a large but finite number of alternative hypotheses K_κ we could work with a continuous infinity of hypotheses. Either approach is an approximation to the other, and which one is adopted is largely a matter of taste. The continuous method is more convenient if the infinitesimal calculus is to be employed. (See **6.5**, example (i).)

It may be asked what exactly is meant by K_κ? There is at present no complete answer to this question, but fortunately this does not appear to matter much. K_κ may be imagined to be the proposition that the man has some particular physical characteristics. For example (very crudely), these characteristics may be that the total weight of those parts of his brain that deal with extra-sensory perception is some assigned function of κ. For our purpose, however, it is sufficient to assume merely that K_κ exists. But if K_κ is not described properly how can the necessary judgment concerning its initial probability be obtained? Any answer that may be given to this can be only a suggestion. It has not been claimed that strict rules can be provided for

---

† Nothing in this book is deliberately directed either for or against a belief in "ESP". In the above work it is assumed that conscious or unconscious cheating is definitely ruled out. An alternative to this somewhat far-fetched assumption is to redefine K as "the man has extra-sensory perception or else there is conscious or unconscious cheating".

deciding on reasonable bodies of belief. But if you take a very long series of trials, you may hope to arrive at a fairly objective view on whether the man has " ESP ", provided that the initial probability judgments are not too prejudiced. Prejudiced initial judgments may be partially avoided by using suggestion (iii) of 4.3. Another suggestion † is that it would be unnatural to take the initial probabilities of say K_{72} and K_{73} as wildly different from each other. To do so would imply that you had a very detailed knowledge of the exact mechanism of ESP. (Cf. the remarks on " smoothness " in 7.5.)

A similar treatment could be provided for testing the amount of bias on a coin. Here it would not be quite so difficult to define the propositions K_κ in detail (provided that a system of dynamical principles was assumed). The difficulty is of the same type as that of defining U in the discussion of scientific theories.

The ideas used in the above example can be applied to any type of experiment in which the probabilities of the possible outcomes depend on the unknown state of some organism or process. Examples are the effect of vaccination of rats, the measurement of intelligence of children, and the quality control of industrial products.

There is one more point that arises in connexion with the example on ESP. In order to make the assumptions correspond more closely with the way in which it is natural to think, it would be necessary to admit the possibility that the " amount of extra-sensory perception " could vary from one trial to the next. This would mean that κ would vary throughout the sequence of trials. For example, it could be held that κ would decrease when the percipient became tired. In order to take this into account, κ would have to be regarded as a function of n, and the probabilities of success at the various trials could be represented by $\frac{1}{2}(1 + \kappa_n/k)$ where $n = 1, 2, \ldots, 1000$. The proposition $\bar{K}$ would be the assertion that $\kappa_1 = \kappa_2 = \ldots = \kappa_{1000} = 0$, and K would be the disjunction of all other possibilities. $\mathcal{B}$ would consist of a set of inequalities for the initial probabilities of every possible sequence $\kappa_1, \kappa_2, \ldots, \kappa_{1000}$. To write out $\mathcal{B}$ in detail would be impracticable, and in fact it would be necessary to be slightly dishonest. Actually it may be best to write down some of the inequalities after looking at the results of the experiment. If, for example, the results of the first 500 trials were much better than the last 500, you might consider that it would lead to sufficiently good results to consider sequences like $\kappa, \kappa, \ldots, \kappa, 0, 0, \ldots, 0$. A particular case of this is the assumption made before that $\kappa_1 = \kappa_2 = \ldots = \kappa_{1000} = \kappa$.

This " dishonesty " can be described more leniently as a very deep judgment that the final probability of K would not be changed much if you went to the trouble of writing out $\mathcal{B}$ in detail. Any assertion such as " it is highly probable

† This is really less of a suggestion than a statement of how people actually think.

that one of the propositions K_{10}, K_{11}, . . ., K_{20} is true" must be taken with a pinch of salt.

Analogous remarks apply to other types of experiments. Often a theory is described as probable when what is meant is that it is probably substantially right. It is unusual to give a precise definition of "substantially right".

4.10 Connexions with the frequency theory

Borel's theorem † provides a connexion between the axiomatic approach and the frequency definition. This theorem can be generalised in an important way.

In Borel's theorem it was supposed that the probabilities of success in a sequence of trials were all equal to p. Problems of a similar type are very often encountered where the probability of success at any given trial depends on the results of previous trials. It is convenient to think in terms of the example of the previous section, but we replace the hypotheses K_κ by a continuous infinity of hypotheses $L_p(0 \leqslant p \leqslant 1)$ such that $P(E_n \mid H.L_p) = p$ and such that $E_1, E_2, E_3, \ldots$ are independent given L_p. It is supposed that one of the hypotheses L_p is true, say L_q, where q is initially unknown.‡ Then it follows from Borel's theorem that *the proportion of successes in the first m trials almost certainly tends to q as m tends to infinity.* Let L_{p_1, p_2} be the disjunction of all L_p for which $p_1 \leqslant p \leqslant p_2$. If it is assumed that $P(L_{p_1, p_2} \mid H) > 0$ whenever $0 \leqslant p_1 < p_2 \leqslant 1$, then it can be proved by using Bayes' theorem T6 that *the probability of E_n, given H together with the results of the first $n - 1$ trials, almost certainly tends to q.* (See also **7.2** and **7.3**.) The two italicised statements will be called the "fundamental theorem of probability". It is of course possible to restate them (as in **1.4** (i) or T20) so as to avoid infinite processes.

The theorem is proved only under the assumptions stated. These assumptions may be more vaguely described by saying that the trials are performed "under the same essential conditions". These essential conditions are $H.L_q$.

A knowledge of this theorem generally causes you to judge that the probability, x, of success at the next trial can be estimated approximately as the proportion y, of successes in a long series of trials, without paying much attention to the initial distribution § of the chance. It may seem to be more accurate

---

† See the remarks following T20 in **3.3**.

‡ q may be called the "true chance" of a success. It is easy to see that all but an enumerable number of the hypotheses L_p must be almost impossible. Thus we are allowing almost impossible hypotheses to occur to the right of the vertical stroke. This can be avoided by complicating the above discussion. One method is to avoid the symbols L_p and to work entirely in terms of the symbols L_{p_1, p_2} with $p_1 < p_2$.

§ It is assumed that the reader is familiar with the idea of a probability distribution. A formal definition is given in chapter 5.

to take the initial distribution into account, but this often entails considerable extra work and may not be worth while.

It is quite legitimate to judge directly that $|x - y| < \delta$ where δ is small, provided that this does not contradict other judgments.† This shows how the frequency approach fits into our probability technique. A contradiction of other judgments is most liable to occur when the equally-probable-cases approach is particularly appropriate. For example, suppose that a coin is spun 1000 times and yields as many as 540 heads. Would you then be willing to judge that the probability of a head at the next trial lies between 0·51 and 0·57? A careful discussion of this example would follow the lines of **4.9** and **6.5**, and will be omitted.

Besides the theoretical connexions between different techniques of probability, there is also the practical connexion that adherents of different schools tend to have somewhat similar judgments. But those who accept the frequency approach often refuse to apply the word " probability " to events that cannot be indefinitely repeated. This is really a question of the use of language. Presumably they do undergo states of more or less belief about such events.

4.11 Relation to the objective theory

A theory in which $P(E \mid H)$ always represents an *objective* degree of reasonable belief has been brilliantly expounded by Jeffreys.‡ It may be regarded more or less as a special case of our theory with the various possible bodies of belief replaced by a fixed objective one, $\mathcal{B}^*$. One of the purposes of the more general theory is to avoid the assumption that $\mathcal{B}^*$ exists. Even if $\mathcal{B}^*$ does exist it is still necessary to fall back on subjective judgments in practice. A juryman may estimate the probability of guilt of a prisoner at more than 0·99 without being able to trace back his opinion to the principle of cogent reason.

An objective theory of probability does not make the problems of section **4.9** any easier to answer.

A truly objective theory or technique which could always be applied in practice, may be impossible of attainment. Such a theory might involve an extensive $\mathcal{B}^*$ or possibly a " complete " list of rules and suggestions, so that no $\mathcal{B}$ would be required at all. While this seems to be quite beyond our powers, there does remain the possibility of adopting extra suggestions. Just as the purpose of the theory is to introduce some measure of objectivity into our bodies of beliefs, the purpose of introducing new suggestions would be to increase this objectivity still further. An attempt to do this has been made by Jeffreys

---

† The specification of δ depends quite a lot on who " you " are. Essentially what is required is an honest judgment. The insistence on an exact rule originates in a respect for science together with the misconception that in science there is no room for judgment.

‡ Jeffreys does not use the description " objective ". See **1.4** (iv), first footnote.

4.12   PROBABILITY AND WEIGHING OF EVIDENCE

(1946). In this paper Jeffreys suggests a plausible form of initial probability distributions for a particular class of cases. These distributions are not deducible from his technique, but they have some invariant properties which suggest that they can be accepted without fear of running into contradictions.

The phrase "the probability of E given H" may make it seem that the theory in this book is an objective one. This would be a misunderstanding based on the conventional use of the definite article. There are two reasons why this use is misleading: first because $P(E \mid H)$ may depend on who you are, and second because the numerical value of $P(E \mid H)$ may be "unobservable". (See 4.4.) The position may be summarised as follows :— It sometimes makes the language simpler to talk as if all the relevant probabilities were objective, but this form of language is strictly justified only for tautological probabilities.

In practice there is sometimes so large an accumulation of evidence that the subjective judgments are obscured. This is why many people have thought that subjective judgments play no part at all. Some adherents of objective techniques are now at loggerheads because in small sample work in statistics the rival objective procedures do not lead to identical results. The present theory abandons the attempt to obtain unique results—it leaves a little freedom of choice to the individual.

A new objective theory has been put forward in recent years by Carnap. His theory involves two types of probability, one of which, called "probability$_1$", corresponds to reasonable and objective degrees of belief. Probability$_1$ is explicitly defined for propositions of a particular kind in terms of the language used. Different languages give rise to different probabilities. (See, for example, Tintner, *Journ. Roy. Stat. Soc.*, Ser. B, 1949 or 1950. In this paper further references may be found.) It is conceivable that "you" could design a language so as to make Carnap's theory consistent with the one presented in the present work. All probability judgments would be pushed back into the construction of the language. Something like Carnap's theory would be required if an electronic reasoning machine is ever built.

4.12 Generalisation of ℬ

So far it has been assumed for simplicity that ℬ must be exhibited in a standard form, before it can be combined with the theory of probability. This standard form consists in a set of equalities and inequalities between degrees of belief. But it is found that judgments of other types can very often be made. One such type has been discussed in 4.3 (i) and in 4.6, namely the direct use of numerical probabilities. Another type mentioned in 1.4 (vii) is a judgment that one course of action is preferable to another one. A new and important type is a direct judgment of "weights of evidence". (See Chapter 6.)

There is no reason why judgments of any sort should be prohibited. This leaves a wide scope for intuition. Whatever form of judgment is used it may be expected to become more discriminating with practice.

With this generalised meaning of $\mathcal{B}$, the function of the theory of probability remains the same as before, namely to enlarge $\mathcal{B}$ and to check up on its self-consistency. (Cf. 4.1, rules (iv), (v) and (vi).)

4.13 Degrees of belief concerning mathematical theorems

If E is a mathematical proposition of a type that is either provable or disprovable, then we know that either $P(E) = 1$ or $P(E) = 0$, by T7, cor. (ii), and T9. As a trivial example let E be the proposition that the millionth figure of π is a 7. Then $P(E) = 1$ or 0. But since the calculations have not been carried out it is natural (at any rate for betting purposes) to assert that $P(E)$ is approximately $\frac{1}{10}$. Unfortunately our theory of probability, in common with most other theories, forces us to reject this judgment.

It may be asked whether the theory could be modified in such a way as to allow judgments of this sort. One way of doing this is by replacing axiom A4 by the following alternative axiom :—

A4'. *If you have seen that E and F are equivalent then $P(E \mid H) = P(F \mid H)$ and $P(H \mid E) = P(H \mid F)$.*

The theory can, I think, be developed in much the same way as in Chapter 3, with axiom A4' replacing A4. One effect of this is that when $\mathcal{B}$ gives rise to a contradiction it becomes correct to say "$\mathcal{B}$ is *now* unreasonable" instead of "$\mathcal{B}$ is unreasonable". Similarly T7, cor. (ii), becomes "when you have proved that H^* implies H then $P(H) = 1$", and so on. This procedure should have some appeal to the intuitionist school of mathematicians.

The question of degrees of belief in purely mathematical theorems is not merely of academic interest. Very often in applied mathematics and chess-playing, in order to save time, a theorem is assumed to be true simply because it is considered to be very likely. One example is the common practice of assuming that the nth term s_n of a convergent sequence is close to the limit, merely because s_n, s_{n-1} and s_{n-2} are close together. (This type of assumption is very frequent in the applications of probability itself.) The effect of the modified axiom is therefore to make the technique of probability more widely applicable.

4.14 Development of the judgment by betting

Probability judgments can be sharpened by laying bets at suitable odds. If people always felt obliged to back their opinions when challenged, we would be spared a few of the "certain" predictions that are so freely made.

The Meteorological Office could set a good example by offering odds with their weather forecasts, provided that some practicable way of doing this could be arranged. Non-betting odds are already very roughly conveyed, otherwise the forecasts would be mere conversation about the weather.

CHAPTER 5

PROBABILITY DISTRIBUTIONS

In this chapter a number of familiar ideas of mathematical probability are described.† This is done for the sake of completeness, and in some places in order to show how these ideas fit into the present theory. Most of the proofs will be omitted.

5.1 Random variables and probability distributions

Suppose that an experiment is performed and that it is known in advance that the result of the experiment will be a real number X. If H is the evidence, assumed not to be almost impossible, let

$$F(x) = P(X \leqslant x \mid H).$$

$F(x)$ " exists " for all x, by axiom A1. It is called the (*probability*) *distribution function* of X (given H), and X is called a *random variable*. In order to save writing, the " misleading notation " of **2.6** will be adopted, i.e. H will be taken for granted and omitted. For example, $P(X \leqslant x)$ will mean $P(X \leqslant x \mid H)$. Clearly, by T9, cor. (iii),

$$F(x_2) - F(x_1) = P(x_1 < X \leqslant x_2),$$

so that $F(x)$ is a non-decreasing function of x.

Although $F(x)$ is assumed to exist it will often not be possible to state it with much accuracy. $\mathcal{B}$ may contain a set of inequalities for $P(x_1 \leqslant X \leqslant x_2)$, $P(x_1 < X \leqslant x_2)$ and so on, for various values of x_1 and x_2. These inequalities will provide information about $F(x)$. In any particular case it will be judged,‡ I think, that $P(x - \varepsilon < X < x)$, $P(X \leqslant -K)$, $P(X > K)$ can be made arbitrarily small by choosing ε sufficiently small and K sufficiently large. If so, it follows at once that

$$\lim_{x \to \infty} F(x) = 1 \quad \lim_{x \to -\infty} F(x) = 0,$$
$$P(X = x) = \lim_{\varepsilon \to 0} \{F(x) - F(x - \varepsilon)\} = F(x) - F(x - 0).$$

The last relation enables us to write down in terms of F the probability that X belongs to any interval of values of x. For example,

$$P(x_1 \leqslant X < x_2) = P(X = x_1) + P(x_1 < X \leqslant x_2) - P(X = x_2)$$
$$= F(x_2 - 0) - F(x_1 - 0).$$

Suppose that X is a physical measurement obtained by reading a scale. It will then be known to lie in a finite interval and will be capable of taking only a

---

† Anyone interested in the advanced mathematical theory should consult Cramér, 1947.
‡ These judgments would not be required if the axiom of complete additivity were assumed.

finite number of values, corresponding to the divisions of the scale. The results $\lim_{\varepsilon \to 0} P(x - \varepsilon < X < x) = 0$, $\lim_{K \to \infty} P(X \leqslant - K) = 0$, $\lim_{K \to \infty} P(X > K) = 0$, will then be forced by T9. Nearly all variables that occur in practice take only a finite number of values; but the notions of infinity and continuity are convenient, since they make available the methods of analysis. Of course, scale readings are often approximations in the sense that greater accuracy could be obtained, but whether they are approximations to variables which are "really" continuous is unanswerable.

It is often convenient to think of F as a differentiable function with derivative $f(x)$, and then $f(x)$ is called the (*probability*) *density* (*function*) of the random variable X. If f exists it is a non-negative function, and assuming only that it is integrable in every finite range, it has the property $\int_{-\infty}^{\infty} f(x)\,dx = 1$. The function $P(X = x)$ is called the (*probability*) *point function of* X. It is suitable for determining the distribution function when the random variable is capable of taking only a discrete set of values (e.g. all the integers).

Let X and Y be two random variables. $P\{(X \leqslant x).(Y \leqslant y)\}$ is called the *distribution function* of the pair of random variables X, Y. Denote it by $F(x, y)$. This may be called a two-dimensional distribution function. The most appropriate mathematical tool for dealing with the general theory of such functions is the two-dimensional Lebesgue-Stieltjes integral.† If the reader is not familiar with this he may be satisfied with accepting the next few remarks in a formal spirit.

Let $Z = \zeta(X, Y)$ be a known function of X and Y. It will have the distribution function $\iint_{\zeta(x,y) \leqslant z} dF(x, y)$. In particular the distribution function of the sum $X + Y$ is $\iint_{x+y \leqslant z} dF(x, y)$.

X and Y are called independent random variables if for all x, and y, the "events" $X \leqslant x$ and $Y \leqslant y$ are independent (at any rate when neither event is almost impossible). Then, by T1, $F(x, y) = F(x)G(y)$, where F and G are the distribution functions of X and Y separately. In particular the distribution function of the sum of two independent random variables is

$$\iint_{x+y \leqslant z} dF(x)\,dG(y)$$
$$= \int_{-\infty}^{\infty} F(z - y)\,dG(y) = \int_{-\infty}^{\infty} G(z - x)\,dF(x).$$

† See, for example, Cramér, 1937.

5.2   PROBABILITY AND WEIGHING OF EVIDENCE

This function will be called the *convolution* of F and G. If F and G are differentiable, the density function of $X + Y$ is

$$\int_{-\infty}^{\infty} f(x-y)g(y)\,dy = \int_{-\infty}^{\infty} g(z-y)f(y)\,dy,$$

a function which is called the *Faltung* or *resultant* of f and g.

5.2   Expectation

If X is a random variable with distribution function F, and if $\psi(x)$ is an arbitrary function of x, then $\int_{-\infty}^{\infty} \psi(x)\,dF(x)$ is called the (mathematical) expectation or expected value of ψ (with respect to the random variable X), assuming of course that this integral exists. It is denoted by $E(\psi)$ or $E(\psi(X))$. In particular suppose that F is differentiable everywhere and that f is the density function. Then

$$E(\psi) = \int_{-\infty}^{\infty} \psi(x)f(x)\,dx.$$

On the other hand, if X can take only a discrete set of values $x_1, x_2, x_3, \ldots$ and if f is the point function, then $E(\psi) = \sum_r \psi(x_r) f(x_r)$.

The expected value of ψ is not necessarily a value that the function can equal. A partial justification for the name "expected value" is to be found in the following theorem, which will not be proved here.

T21   *If $X_1, X_2, X_3, \ldots$ are independent random variables, all with the same distribution function, then it is almost certain that*

$$\frac{1}{n}(X_1 + X_2 + \ldots + X_n) \to E(X_1) \text{ as } n \to \infty.$$

Borel's theorem, equivalent to T20, is the special case of this in which the random variable is 1 or 0 according as a "trial" is successful or unsuccessful.

A more general theorem than T21 is the following.

T21A   *If $X_1, X_2, X_3, \ldots$ are independent random variables for which $E(X_r^2)$ is bounded, then it is almost certain that*

$$\frac{1}{n}(X_1 + X_2 + \ldots + X_n) - \frac{1}{n}\{E(X_1) + E(X_2) + \ldots + E(X_n)\} \to 0$$

as $n \to \infty$.

COROLLARY. *In particular the conclusion applies if all the random variables are restricted to a fixed finite interval.*[†]

Suppose that an experiment with result X is followed by a monetary gain

---

[†] T21A is equivalent to a special case of the so-called strong law of large numbers, itself generalised in an interesting manner by Kolmogoroff and Khintchine. For an excellent introductory account of these and other generalisations see Feller, 1945.

of amount $\psi(X)$. Then $E(\psi)$ is called the *expected monetary benefit* (of the experiment). Similarly the expected gain of "utility" can be defined. "Utility" is the economist's name for a "reasonable" measure of "value".† Utilities may sometimes be subjectively compared in the same way as probabilities. A utility is best regarded as depending on a "change of circumstances". This is not a concept that belongs to classical logic, so that it would hardly be possible to build up an abstract theory of utility. But the analogues of the "obvious axioms" of 2.2 could hardly be disputed. These can be extended, just as for probabilities, by assuming that a utility is a real number that vanishes when there are no changes of circumstances. In order to obtain results of interest it is necessary to be able to judge the numerical value of a ratio of two utilities. This ratio need be judged merely to lie in some interval, possibly a very wide one.

In virtue of T21 and T21A it is rational to behave in such a manner as to maximise the expected utility. In this way any theory of probability can be taken as a guide to action. Perhaps all practical applications of probability can be regarded from this point of view. In fact, as mentioned in 1.4 (vii), Ramsey takes expected utility as a primitive notion and defines degrees of belief in terms of it. It seems simpler and more natural to treat beliefs and values as distinct subjective notions, but the direct judgment of expected utilities is permissible in the generalised form of our theory (see 4.12).

An insurance company is willing to regard the utility of a monetary gain or loss as proportional to the amount of money. This would not be true for amounts that were large compared with the total capital of the company. Since insurance companies usually have very large capitals, actuaries can work directly with expected monetary benefits.

It seems rational to assume that as a general rule the utility of money is a concave function of the total capital, when this is positive. A consequence is that it is not worth taking a level bet if the probability of winning is only $\frac{1}{2}$. On the other hand an insurance policy can very well provide a positive expected utility in spite of a negative expected monetary benefit. This remark applies even to life insurance, for reasons that the reader can think out for himself. Another example of expected utilities is provided by the "Petersburg problem".

> "A coin is spun an indefinite number of times and if there is a run of n heads before the first tail there is a prize of 2^{n+1} units. How much should be paid for the privilege of playing?"

Worked out in terms of expected monetary benefit the result is infinite. A

† This "value" depends on ethics and on amounts of happiness. The distinction between utility for an individual and utility for a group of individuals will not be discussed here.

finite value for the expected utility can be obtained by assuming that the utility of a sum of money is proportional to the logarithm of the amount measured in suitable units, as suggested by Daniel Bernoulli. (See Todhunter (1865), 220.) This assumption is inadequate since it would still lead to an infinite result for a slightly modified game, in which the amount 2^{n+1} is replaced by 2^{2n+1}. In order to get a finite result for all such modifications it must be assumed that there is an upper bound for the amount of utility of money, where the upper bound may depend on the individual. If, for example, the utility is a concave function of the amount and if this function is constant for amounts of more than 2^{20} units, then the game is not worth more than 21 units. The proof is left to the reader. The entrance fee that is worth paying for n games is not necessarily equal to n times that for one game. (We have throughout disregarded the utility of gambling itself.)

Suppose that it is assumed quite generally that utilities are bounded. Then T21A cor., when expressed in a finite form (without the use of limiting processes), can be used to provide a fairly complete justification of the principle of maximising expected utilities.

The idea of mathematical expectation is continually used in the study of probability distributions. Examples are (i) the *moments* $E(X^r) = \mu'_r$ ($r = 0, 1, 2, \ldots$), where $\mu'_0 = 1$, and μ'_1 is the *mean (value)* of X, (ii) the *moments about the mean*, $E\{(X - \mu'_1)^r\} = \mu_r$, where $\mu_0 = 1, \mu_1 = 0, \mu_2 =$ the *variance* $= \sigma^2$ where $\sigma \geqslant 0$ and is called the *standard deviation*, (iii) the *characteristic function* $E(e^{iXt})$. Unlike X, t is an ordinary mathematical variable. The integral for the characteristic function always converges, but those for the moments may not all converge. Under fairly general conditions a distribution is determined by a knowledge of all the moments or of the characteristic function.

In fact if the characteristic function is $\varphi(t)$, then the point function at x is

$$p(x) = \lim_{T \to \infty} \frac{1}{2T} \int_{-T}^{T} \varphi(t) e^{-ixt} dt,$$

and F can then be determined from

$$F(x_2) - F(x_1 = \tfrac{1}{2}\{p(x_2) - p(x_1)\} + \lim_{T \to \infty} \frac{1}{2\pi} \int_{-T}^{T} \varphi(t) dt \int_{x_1}^{x_2} e^{-ixt} dx \, ;$$

while at a point x at which there is a density function, it is

$$f(x) = \lim_{T \to \infty} \frac{1}{2\pi} \int_{-T}^{T} \varphi(t) e^{-ixt} dt.$$

The moments may be formally deduced from the characteristic function by expanding the exponential and integrating term by term. *The characteristic function of a convolution of two distributions is the product of the separate characteristic functions.*

The mean and standard deviation are good measures of the " typical value " and " spread " of a distribution. There are other such measures, such as the median value, μ, for which $F(\mu) = \frac{1}{2}$, and the mean deviation $E(\,|\,X - \mu_1'\,|\,)$. These have some advantages for numerical work but are more difficult to deal with in the mathematical theory.

5.3 Examples of distributions

Suppose that a random variable X_1 is known to lie strictly between two numbers a and b. It is sometimes said that if nothing more is known about X, then its density function must be $\dfrac{1}{b-a}$, i.e. constant throughout the interval (a, b). The distribution is said to be rectangular or uniform (cf. **2.8**). This is essentially an application of the principle of insufficient reason, or of " Bayes' postulate " (rather than " Bayes' theorem "). But in practice there is always some additional information about X, and the uniform distribution occurs only as an approximation. We should sometimes judge that for some specified constant $\lambda > 1$,
$$P(x_1 < X < x_2) > P(x_1' < X < x_2')$$
whenever
$$x_2 - x_1 > \lambda(x_2' - x_1'), \quad a < x_1 < x_2 < b, \quad a < x_1' < x_2' < b.$$
If λ is close to 1 the numerical consequences of adopting these judgments would be much the same as if Bayes' postulate had been accepted.

The standard type of argument against Bayes' postulate is that if all that is known about X is that it lies between a and b, then all that is known about, say, X^{100} is that it lies between a^{100} and b^{100}; and Bayes' postulate applied to the random variables X and X^{100} gives two quite different distributions for X. Fortunately Bayes' postulate is not required in the present theory. For if X arose in a fairly natural way, say as a volume, it would be entirely artificial to introduce the random variable X^{100}. You would simply not judge honestly that the distribution of X^{100} was anything like uniform.

Next suppose that X is known to lie in a *closed* interval, i.e. $a \leqslant X \leqslant b$. It was proposed by J. B. S. Haldane and H. Jeffreys † that if nothing more is known, then a finite amount of the probability must be concentrated at a and b. This shows how distributions can arise that are neither continuous nor discrete.

If X is known only to be a real number, the assumption of a uniform distribution forces the use of infinite probability to represent certainty, with an appropriate modification of the axioms. A reference to this has already been

---

† See Jeffreys, 1939, 114 and Haldane, 1931. It would be quite rational to concentrate a finite amount of probability at every " computable " value of x, the largest amounts being concentrated at the simplest values. (Cf. **5.4**.) It is possible to imagine this done since the computable numbers form an enumerable set.

5.3 PROBABILITY AND WEIGHING OF EVIDENCE

made in 3.1 (ix). Similarly, if X is known to be positive, Haldane and Jeffreys assume a uniform distribution for $\log X$, i.e. a density function $\frac{1}{x}$ for X. This also involves infinite probabilities. In both these cases the use of infinite probability can be avoided in practice by using known bounds for x (which always exist). In the second case, one of the bounds is some small positive number, and it may very well be judged that the distribution of $\log X$ is approximately uniform over a *finite* range.

Three distributions which occur a great deal, as approximations † at least, in practical and theoretical work, are the binomial, the Poisson and the normal distributions. The first two are discrete distributions and have point functions

$$P(X = r) = \binom{n}{r} p^r (1-p)^{n-r} \quad (r = 0, 1, 2, \ldots n\,;\ 0 < p < 1),$$

and $P(X = r) = e^{-a} a^r / r!$ $\quad (r = 0, 1, 2, \ldots\,;\ a > 0).$

The first of these was mentioned in T19. The normal distribution has density function

$$\frac{1}{\sigma\sqrt{2\pi}} e^{-\frac{1}{2\sigma^2}(x-x_0)^2}.$$

The corresponding characteristic functions are respectively

$$(e^{it} p + 1 - p)^n, \quad \exp\{a(e^{it} - 1)\}, \quad \exp(x_0 ti - \tfrac{1}{2} t^2 \sigma^2).$$

From these the moments may be deduced. In particular the means are pn, a, x_0 and the standard deviations are $\sqrt{np(1-p)}$, $\sqrt{a}$, σ. Another deduction from the form of the characteristic functions is that the convolution of a number of Poisson distributions is again a Poisson distribution, with a similar result for normal distributions.

If $n \to \infty$ and $p \to 0$ in such a way that $pn = a$, a constant, then the first characteristic function tends to the second one. This suggests (correctly) that the point function for the binomial distribution may be approximated by that for the Poisson distribution if n is large but pn is moderate.

If a distribution with characteristic function $\varphi(t)$ is expressed in terms of a new variable $(x - \mu_1')/\sigma$ it is said to be expressed in standard measure. In terms of the new variable the mean is 0 and the standard deviation is 1. The new characteristic function is $e^{-it\mu_1'/\sigma} \varphi\!\left(\dfrac{t}{\sigma}\right)$. If the binomial, Poisson and normal distributions are expressed in standard measure, the corresponding characteristic functions of the first two tend to the last one. Hence it is not

---

† A natural way of expressing the order of the approximation is by giving upper and lower bounds for the proportional error at each value of x for the point or density function, or in each interval (x_1, x_2) for $P(x_1 < X < x_2)$. Cf. the first paragraph of 5.3.

PROBABILITY DISTRIBUTIONS 5.3

surprising that the distributions themselves, in standard measure, tend † to $\frac{1}{\sqrt{2\pi}}e^{-\frac{1}{2}x^2}$. This is a special case of a result called the *central limit theorem,* which states that under rather general conditions, the convolution, when expressed in standard measure, of n independent distributions tends to $\frac{1}{\sqrt{2\pi}}\int_{-\infty}^{x} e^{-\frac{1}{2}t^2}dt$. (See also Appendix I.)

For a very much fuller discussion of the theory of general and special distributions the reader is referred to Kendall (1945), Wilks (1944), or Cramér (1946).

Exercises

(i) Prove Tchebycheff's inequality, that
$$P(|x - \mu_1'| > \lambda\sigma) \leqslant \lambda^{-2},$$
whatever the distribution function.

(ii) A random variable X has a density function $f(x)$, which is continuous for all x. Let ξ_r be the rth digit of the fractional part of X when X is expressed as an infinite decimal. Show that $P(\xi_r = 7) \to 0.1$ as $r \to \infty$. (Hint: assume first that $f(x)$ vanishes outside a finite interval and prove
$$P(\xi_r = 6) - P(\xi_r = 7) \to 0, \text{ etc.})$$

(iii) A well-balanced wheel can be spun rapidly about its centre. The wheel is divided into 10 equal sectors numbered 0, 1, 2, ..., 9. (Cf. Kendall (1945), 189.) The wheel is spun, starting from a known position, and is allowed to rotate for a time. The number of revolutions of the wheel is a random variable. The digit opposite a fixed pointer at the end of the time is another random variable. Discuss the connexion between this physical experiment and the result of exercise (ii).

(iv) A form of Stirling's formula is
$$\log t! = (t + \tfrac{1}{2})\log t - t + \tfrac{1}{2}\log 2\pi + \frac{\theta(t)}{12t},$$
where $t > 0$, $0 < \theta(t) < 1$. (See, for example, Jeffreys (1939), 371–2.) Using this formula show that
$$\log f(\lambda n, \lambda r) - \lambda \log f(n, r) = \frac{\lambda - 1}{2}\log\frac{2\pi r(n-r)}{n} - \tfrac{1}{2}\log\lambda + \rho,$$
where
$$f(n, r) = \binom{n}{r}p^r(1-p)^{n-r}, \quad \lambda > 1,$$
$$|\rho| < \frac{\lambda + 1}{12}\left(\frac{1}{n} + \frac{1}{r} + \frac{1}{n-r}\right).$$

† This method of approximating the binomial distribution is what was required in the proof of T20.

Hence show that

$$\log \psi(\lambda n, \lambda r) = \lambda \log \psi(n, r) + \frac{\lambda - 1}{2} \log \left\{ \left(1 + \frac{x\delta}{pn}\right)\left(1 - \frac{x\sigma}{(1-p)n}\right) \right\} + \rho,$$

where

$$\psi(n, r) = f(n, r)/g(\sigma, x),$$
$$g(\sigma, n) = \frac{1}{\sigma\sqrt{2\pi}} e^{-\frac{x^2}{2\sigma^2}}, \quad \sigma = \sqrt{np(1-p)}, \quad x = \frac{r - pn}{\sigma}.$$

If $p = \frac{1}{2}$ show that $\psi(5000, 3250)$ is about 0.027, given that $\log_{10} \psi(100, 65) = -0.0112$.

(v) A sequence of digits each have *chances* $p_0, p_1, \ldots, p_9$ of being 0, 1, ..., 9. These digits are added " modulo 10 " in blocks of N, thus producing a new sequence with chances $p'_0, p'_1, \ldots, p'_9$. Show that

$$p'_r = \frac{1}{10} \sum_{s=0}^{9} \{\varphi(s)\}^N \omega^{-rs},$$

where

$$\omega = e^{2\pi i/10} \quad \text{and} \quad \varphi(s) = \sum_{r=0}^{9} p_r \omega^{rs}.$$

(Hint: first prove the special case † $N = 1$ and find a result analogous to the multiplicative property of the characteristic function of the sum of independent random variables.)

Deduce that

$$10 \sum_{r=0}^{9} (p'_r - \tfrac{1}{10})^2 = \sum_{s=1}^{9} |\varphi(s)|^{2N} \leqslant 9\mu^{2N},$$

where $\mu = \text{aver} \,|\, 10 p_r - 1 \,|$.

(vi) X and Y are a pair of random variables with distribution function

$$F(x, y) = \int_{-\infty}^{x} \int_{-\infty}^{y} f(t, u) \, dt \, du.$$

The expectation of a function $\psi(X, Y)$ is defined as

$$E(\psi(X, Y)) = \int_{-\infty}^{\infty} \int_{-\infty}^{\infty} \psi(x, y) f(x, y) \, dx \, dy.$$

Let the analogues of inertial constants of a rigid body be defined by the equations

$$\mu'_1 = E(x), \quad \nu'_1 = E(y), \quad \sigma^2 = E\{(x - \mu'_1)^2\},$$
$$\tau^2 = E\{(y - \nu'_1)^2\}, \quad \sigma\tau\rho = E\{(x - \mu'_1)(y - \nu'_1)\}.$$

(ρ is called the *correlation coefficient* between X and Y.) Show that the variance of $X + Y$ is $\sigma^2 + \tau^2 + 2\sigma\tau\rho$. Show that the probability density of X alone exists and equals $\int_{-\infty}^{\infty} f(x, y) \, dy$.

† Cf. Weyl, *The theory of groups and quantum mechanics* (London, 1931), 34.

(vii) Let $\varphi(t)$ be a characteristic function of a distribution and let $\log \varphi(t) = \sum_{r=1}^{\infty} \kappa_r \frac{(it)^r}{r!}$, assuming such an expansion is permissible. $\kappa_1, \kappa_2, \kappa_3,$ ... are called the *cumulants* of the distribution. Show, at any rate formally, that the cumulants for the sum of independent variables are equal to the sums of the corresponding cumulants.

(viii) Prove that $\mu_1' = \kappa_1$, $\mu_2 = \kappa_2$, $\mu_3 = \kappa_3$, $\mu_4 = \kappa_4 + 3\kappa_2^2$. Hence show that the mean, the variance and the third moment about the mean for the sum of any number of independent variables are equal to the sums of the individual means, variances and third moments about the mean. The first of the three results is true also for variables that are not independent. The second part may be compared with exercise (vi).

5.4 Statistical populations and frequency distributions

Imagine that the heights are known to the nearest inch of all the men in England. Let $\varphi(r)$ be the number of men of height r inches. Let $N = \sum_{r=0}^{\infty} \varphi(r)$, the total number of men. Let $f(r) = \varphi(r)/N$. Let $F(x) = \sum_{s \leqslant x} f(s)$. Then $F(x)$ is called the *frequency distribution function* of r. It is defined without reference to probability, but it is equal to the probability distribution function associated with the experiment of selecting men *at random* from the population. (See 4.7 (i).) The obvious name for $f(r)$ is the "(frequency) point function". The mean, variance, etc. can be defined in the same way as for general distribution functions. If the population is regarded as large and the "class interval" (one inch) as small, then it may be convenient to approximate to $F(x)$ by a differentiable function of the height and to introduce a density function.

The usual statistical method of finding out properties of a "population" is to take only a partial sample. This is more convenient than examining the whole population. When the population is virtually infinite, as in dice-throwing, it is impracticable to take more than a partial sample. The partial sample can itself be regarded as a population,† and it will have its own frequency distribution which can always be described without introducing probabilities. But it would be useful to be able to deduce that the frequency distribution of another sample would be approximately the same, provided that both samples were reasonably large. No such deduction is possible without using the ideas

† But this word is usually reserved for the whole population from which the sample is drawn.

of probability. This explains an essential connexion between statistics and probability. The question will be discussed again in the last chapter.

When a sample is regarded as a population, with a frequency function, the mean, variance, etc. of this function are called the sample mean, sample variance, etc. These have some relation to the mean, variance, etc. of the whole population, but should not be confused with them.

When a frequency distribution is obtained from statistics, there is no particular reason to suppose that it is expressible in a simple mathematical form. But it is often possible to find a simple form that fits the frequency distribution approximately. If this can be done it has the advantage of describing the results of the statistics briefly. In some cases it is suggestive of the causes that lie behind the results. But the main reason, in general, for looking for a simple mathematical " law " † of this type is that if it is found it is believed to have predictive value. That is to say the simple law, if it is a very good approximation to the distribution function F of the original sample, is likely to describe the distribution function of another sample (or of the whole population) *even better than F would*. This is partly because it is likely that there are a few predominating causes lying behind the statistics, even though these causes are unknown. ‡ If there are such causes then it is natural to suppose that any given simple law has a non-negligible initial probability of being a good approximation. This probability will change when the statistics are taken into account, and may become close to one if the sample is not too small. It will be realised that these remarks are not intended to be precise. They are in the nature of " suggestions ". They are a special case of the general principle of simplicity known as " Occam's razor ". (See, for example, Jeffreys (1939), 277.)

One of the difficulties is how to decide on initial probabilities of laws. No simple complete suggestions can be given, if only because it often happens in statistical experiments that similar experiments have been done before and this complicates the initial evidence a great deal. In particular the normal law is often favoured because it is known to have occurred approximately § in previous experiments, and because it is easy to treat mathematically.

A plausible formula for the initial probability of a law containing n parameters is 2^{-n}, provided that there is no initial evidence at all. (See Jeffreys

---

† In the remainder of this section the word " law " refers to the frequency distribution in the whole finite population. Most of the remarks would apply, with a little modification, to " hypothetical infinite populations " (see **7.2**) and also to scientific laws in general.

‡ It is by no means necessary for the simplicity of a law that the number of predominating causes should be small.

§ The approximation often becomes rather poor, as a percentage, in the " tails " of the distribution, i.e. at more than a few σ from the mean. (Cf. **5.3**, exercise (iv).)

(1939), 96.) An objection to this is that there are several laws of different forms with the same number of parameters. It seems therefore that in the present state of the theory something must be left to the individual judgment. As regards the initial distribution of the parameters, once the form of the law has been decided, it may be natural to assume in some cases that the parameters or their logarithms are approximately uniformly distributed.

The general problem of specifying probability distributions of frequency distributions can be expressed in terms of the measurement of volume in a " space of functions ". The problem is a difficult one if the number of parameters in the frequency distributions is infinite.

CHAPTER 6

WEIGHING EVIDENCE

"Mathematical reasoning and deductions are a fine preparation for investigating the abstruse speculations of the law." THOMAS JEFFERSON

6.1 Factors and likelihoods

The main purpose of the present chapter is to provide a quantitative description of the ordinary process of weighing evidence.† The discussion is closely connected with Section 4.9, being based on Bayes' theorem T6. If in that theorem H is taken for granted, as in Chapter 5, it may be written

$$\frac{P(E \mid F)}{P(E)} \propto P(F \mid E),$$

or after a change of notation,

$$\frac{P(H \mid E)}{P(H)} \propto P(E \mid H),$$

where E is fixed and H is variable. The reason for the new notation is that for most of the applications H is considered as a hypothesis and E as (the proposition asserting) the result of an experiment. The theorem is known also as the *principle of inverse probability*.

$P(E \mid H)$ may be called the *likelihood* of H given E. The term was introduced by R. A. Fisher with the object of *avoiding* the use of Bayes' theorem.‡

The theorem may be expressed " The ratios of the final to the initial probabilities of a set of hypotheses are proportional to their likelihoods ".

The simplest case is when there are only two hypotheses, which may then be represented by H and $\bar{H}$. We then find that

$$\frac{O(H \mid E)}{O(H)} = \frac{P(E \mid H)}{P(E \mid \bar{H})},$$

where $O(H \mid E)$ is defined as $P(H \mid E)/\{1 - P(H \mid E)\}$, and is called the *odds* of H given E. It is natural to call $O(H)$ the initial odds and $O(H \mid E)$ the final odds. In general, if p is any probability, the corresponding odds are defined as $o = p/(1-p)$, so that $p = o/(1+o)$. If $o = m/n$ it is often said that the odds are " m to n on " or " n to m against ". These should not be confused with betting odds. Odds of 1 are called " evens ".

---

† A non-mathematical discussion of the subject is given in chapters XVI and XVII of Venn, 1888.
‡ See **7.1**, **7.4** and Fisher, 1938, 11 and 15.

WEIGHING EVIDENCE 6.1

$O(H \mid E)/O(H)$ is the factor by which the initial odds of H must be multiplied in order to obtain the final odds. Dr. A. M. Turing suggested in a conversation in 1940 that the word "factor" should be regarded as a technical term in this connexion, and that it could be more fully described as *the factor in favour of the hypothesis H in virtue of the result of the experiment.*

The ratio $P(E \mid H)/P(E \mid \bar{H})$ is the ratio of the likelihoods † of H and $\bar{H}$ with respect to E. The particular case of Bayes' theorem may accordingly be stated as

T22 *The factor in favour of a hypothesis H is equal to the ratio of the likelihoods of H and $\bar{H}$.*

Because of this theorem the word "factor" will be used indiscriminately for $O(H \mid E)/O(H)$ and for the ratio of the likelihoods. The reason for preferring the word "factor" is that it is from our point of view the practical significance of the ratio of the likelihoods. The factor in favour of a hypothesis is equal to the final odds when the initial odds are evens. (It is therefore equal to the number that Jeffreys denotes by "K".)

Turing suggested further that it would be convenient to take over from acoustics and electrical engineering the notation of bels and decibels (db). In acoustics, for example, the bel is the logarithm to base 10 of the ratio of two intensities of sound. Similarly, if f is the factor in favour of a hypothesis, i.e. the ratio of its final odds to its initial odds, then we say that the hypothesis has gained $\log_{10} f$ bels ‡ or $(10 \log_{10} f)$ db. This may also be described as the *weight of evidence* § or amount of information ‖ for H given E, and $(10 \log_{10} o)$ db may be called the *plausibility* ¶ corresponding to odds o. Thus T22 may be expressed :

"Plausibility gained = weight of evidence",

where the weight of evidence is calculated in terms of the ratio of the likelihoods.

The use of the words "factor", "decibel" etc. receives particular significance from the following simple theorem.

T23 *Suppose that a series of experiments are performed, with results E_1,*

---

† The phrase "likelihood ratio" is sometimes reserved, in statistical literature, for the expression x/x', x and x' being the maxima of $P(E \mid H)$ when H runs through two sets, S and S', of hypotheses, S being a subset of S'.

‡ "Natural bels" can be defined in a similar way by using natural logarithms instead of common logarithms. A natural bel is then 4·343 db. In electrical engineering a "neper" is 8·686 db.

§ In 1936 Jeffreys had already appreciated the importance of the logarithm of the factor and had suggested for it the name "support". (See *References*.)

‖ The phrase "amount of information" is used in a different sense by Fisher. (For yet another sense see **6.9**.)

¶ The use of the term "plausibility" in very nearly this way was suggested by Professor J. B. S. Haldane, after he had kindly read a draft of the present chapter. He suggests an "octave" for the weight of evidence corresponding to a factor of 2. I am much indebted to him for some useful criticisms.

$E_2, \ldots, E_n$ and suppose that these are independent given H and independent given $\bar{H}$. Then the resulting factor is equal to the product of the individual factors, and therefore the resulting weight of evidence is equal to the sum of the individual weights of evidence.

For

$$\frac{P(E_1.E_2.\ldots E_n \mid H)}{P(E_1.E_2.\ldots E_n \mid \bar{H})} = \frac{P(E_1 \mid H)}{P(E_1 \mid \bar{H})} \cdots \frac{P(E_n \mid H)}{P(E_n \mid \bar{H})},$$

because of the independence conditions, so that factors are multiplicative and weights of evidence are additive.

Example. A die is selected at random from a hat containing ten homogeneous dice and one loaded one. The loaded one is assumed to have a *chance* of $\frac{1}{3}$ of yielding a 6. The selected die is thrown nine times and comes down 6 eight times. What are the final odds that it is the loaded one?

The initial plausibility for the selected die's being loaded is $10 \log_{10} \frac{1}{10}$ $= -10$ db. For each 6 the hypothesis gains a factor of $\frac{1}{3}/\frac{1}{6}$, i.e. very nearly 3 db since $\log_{10} 2 = 0.301$. For each non-six it loses a factor of $\frac{5}{6}/\frac{2}{3}$, i.e. nearly 1 db. Hence the net gain is 23 db, the final plausibility is 13 db, and the final odds are 20 (or " 20 to 1 on ").

This example shows that the decibels used here and those used in acoustics and electrical engineering have similar advantages for mental work.

The decibel might be defined quite generally as ten times the logarithm to base 10 of a ratio. It may be convenient in other connexions, apart from the theory of probability, acoustics and transmission lines. For example, the ratio of brightness of two stars differing by one magnitude is exactly 4 db. The frequency ratio corresponding to a semitone in music is very close to $\frac{1}{4}$ db, since there are twelve semitones in an octave.

6.2 " Sequential tests " of statistical hypotheses

In 1943 A. Wald † developed a technique for the quality control of goods and for deciding between two courses of action. The technique was applied in thousands of American factories during the war. The basic idea can be expressed in terms of factors and weights of evidence, although this terminology was not used by Wald.

Suppose that some article is produced in wholesale quantities. The whole collection of articles is called the " lot " and is supposed to be very numerous. Some of the articles are selected at random, one by one, and put to some test. E represents the proposition that one article passes the test. There are two hypotheses H and $\bar{H}$ concerning the articles. These two hypotheses are such that $P(E \mid H)$, $P(E \mid \bar{H})$ have assigned values and are *chances*. An alternative approach would be to define H and $\bar{H}$ as stating that two fixed proportions of

† See Wald, 1945 (two references) or 1947 and Barnard, 1946.

the lot would pass the test. This approach would lead to nearly the same results if the lot were assumed to be large compared with the sample.

The object of testing the goods is to decide between H and $\bar{H}$. (The case of more than two hypotheses will not be discussed here.) It may be too expensive to test all the articles in the lot; for example, the test may be a destructive one.

Whenever an article passes the test, the hypothesis H has a plausibility gain of
$$10 \log_{10} P(E \mid H) - 10 \log_{10} P(E \mid \bar{H}) \text{ db.}$$
When an article fails to pass the test there is a loss of
$$10 \log_{10} \{1 - P(E \mid \bar{H})\} - 10 \log_{10} \{1 - P(E \mid H)\} \text{ db.}$$
Before the testing is begun a decision should be made as to how much plausibility should be gained or lost by H before the lot is accepted or rejected. The testing need be continued only until one of the levels is reached. This means that the number of tests cannot be predicted, but the *expected* number required is naturally less than if the method depended on a sample of fixed size. The technique is very easy to apply once the required levels of plausibility gain and plausibility loss have been decided. The estimation of these levels can be made to depend on estimates, possibly within wide intervals of

(i) the initial odds of H,

(ii) the utility gains and losses involved in accepting H when H is true or false or in rejecting it when true or false,

(iii) the utility loss of one test (or the cost of one test).

Wald's method of deciding on the required levels is different. It depends on estimates of

(iv) the largest number α which can be tolerated for the probability of rejecting H when H is true,

(v) the largest number β which can be tolerated for the probability of accepting H when H is false.†

Wald is quite aware of the connexion of his technique with Bayes' theorem, but he adopts the second method of estimating the required weights of evidence because of the desire to use only objective probabilities. Our contention is that the judgment of α and β is just as subjective as the judgment of $O(H)$. Wald's method is easier to apply once the subjective judgments are made.

When α and β are given, Wald proves that the technique leads approximately to a smaller expected number of tests than any other technique, whether H is true or false. This result is hardly surprising since the factor obtained from

† See the definitions of "errors of the first and second kinds" in 7.4.

the whole experiment tells us as much about the probability of H as it is possible to deduce from the experiment. (See also **6.7**.)

The sequential technique is clearly not restricted to the quality control of goods. It can be used for deciding between any two " simple statistical hypotheses " (in a sense to be defined in **7.4**).

6.3 Three hypotheses and legal applications

When there are three possible hypotheses H, H' and H'', it may still be convenient to consider them in pairs. For example, it may be decided in the first place to ignore H'', i.e. to take $\bar{H}''$ for granted. In order to simplify the notation, $\bar{H}''$ may be absorbed into the " vague general information " that is left out of account. It then becomes only slightly misleading to denote H' by $\bar{H}$, and the language of odds, factors etc. becomes available. If in this way the evidence is such as to decide " definitely " between H and H', then H'' may be reintroduced. There will again be only two hypotheses to take into consideration and the technique for two hypotheses may be applied again.

This method corresponds to a natural way of thinking about legal cases. There are often three hypotheses that are worth distinguishing: that the evidence is fortuitous,† that a particular man is guilty, or that this man has been " framed ". The last hypothesis will normally be left out of account (together, perhaps, with others) until the choice between the first two hypotheses is fairly clear. Similarly in card-guessing experiments the results might be due to chance, to extra-sensory perception or to conscious or unconscious cheating. Here again the last possibility would often be ignored until the second one had become more plausible than the first.

In general when there are more than two possible hypotheses it is often convenient to " take them for granted " in pairs, so that one of a pair can be regarded as the negation of the other. The method is commonly adopted in statistics and some examples will be given in Chapter 7. In fact a great deal of thinking in statistics, science and ordinary life consists in taking hypotheses for granted in pairs. This often leads ultimately to very high odds for one of the hypotheses, and it then becomes important to remember that there may be other hypotheses to consider.

The technique of decibels may be used in an approximate way for legal purposes. If for example a crime is committed in London, the initial plausibility of guilt of a particular Londoner is roughly — 70 db. Therefore 90 db are needed in order to bring the odds up to 100 to 1 on. The various pieces of evidence (in the ordinary sense) supply different weights of evidence and

---

† We do not mean to imply that no crime was committed at all, but merely that the suspect was involved in a non-causal manner; by happening to be near the scene of the crime, for example.

WEIGHING EVIDENCE 6.3

the results may be added, *if the pieces of evidence are independent*; otherwise some allowance must be made for the degree of dependence. The appropriate number of decibels to be allotted for any piece of evidence would be largely a matter of experience and judgment. It seems likely that the use of decibels in this way would be of considerable value once it had become a mental habit. Many ordinary commonsense ideas would be given a rough numerical basis and would therefore be made clearer. (Cf. **4.12**.)

Consider why it is important to find a motive in a murder case. The reason is that it is much more probable that a man will commit murder with a known motive than without one. The ratio of these probabilities therefore supplies a large factor in favour of guilt. Similarly, in the case of theft, a man with several convictions is more likely to be suspected. The correct factor in favour of guilt in virtue of previous convictions could be obtained approximately by statistical methods. Without the statistics there is a danger that the factor would be overestimated. This is why juries are not supposed to take previous convictions into account. It is perhaps somewhat inconsistent that the *appearance* of the accused man is allowed to influence the jury.

It is convenient to refer here to a principle stated by Sherlock Holmes. *If a hypothesis is initially very improbable but is the only one that explains the facts, then it must be accepted.* From the present point of view this is because the hypothesis receives an infinite factor from the evidence. The principle is often used in scientific work. It is liable, however, to be misleading. For if the only hypothesis that seems to explain the facts has very small initial odds, then this is itself evidence that some alternative hypothesis has been overlooked. This too is an example of Bayes' theorem !

A similar point can be exemplified by means of the hat containing eleven dice, mentioned in **6.1**. Suppose that the selected die had been thrown 60 times. What number of 6's would make it most convincing that the selected die was the loaded one ? Some people would reply that the best number of 6's would be 20 since this is the expected number if the die is known to be loaded. This would be an example of what may be called " the fallacy of typicalness ". In fact the more 6's that are obtained the more probable it is that the loaded die has been selected. But in practice we could never *know* that the hat contained eleven dice of the type mentioned—we could regard it merely as highly probable. Thus, if all 60 throws yielded a 6, we should get $600 \log_{10} 3 = 286$ db in favour of the view that the loaded die had been surreptitiously replaced by a " completely loaded " one, provided that there were no other hypothesis that could be considered. A similar argument arose in connexion with the Dreyfus case, where there was so much circumstantial evidence as to suggest that Dreyfus had been framed.

6.4 Small probabilities in everyday life

In ordinary life you continually use Bayes' theorem in some form. Sometimes the initial probabilities are very small but the factors are very large. For example, if you meet a "random man" in France, the initial probability may easily be as small as 10^{-12} that he is a particular Englishman with whom you are acquainted. But if he happens to be the Englishman in question, it is generally fairly easy to recognise him (though not as easy as when he is in his normal environment). It follows that you can quickly observe enough characteristics of the man so that the probability is less than 10^{-12} that another man, selected at random in France, would have the same characteristics. (For a factor of at least 10^{12} is required.)

6.5 Composite hypotheses

In general, when there are more than two hypotheses, the natural procedure is to work with the original form of Bayes' theorem. But there is a case that is in a sense intermediate between the cases of two hypotheses and of more than two. Suppose in fact that you wish to know whether a hypothesis H is true, the evidence being E (together with some evidence H' which is taken for granted). Suppose further that H can be expressed in a convenient way as the disjunction of n mutually exclusive hypotheses $H_1, H_2, \ldots, H_n$. Then H may be described as a *composite hypothesis*. (See also 7.4.)

If it were assumed that $H_2, H_3, \ldots, H_n$ were false the factor in favour of H in virtue of E would be $P(E \mid H_1)/P(E \mid \bar{H})$. Denote this expression by f_1 and let $f_2, f_3, \ldots, f_n$ be defined in a similar way. These numbers are analogous to the partial derivatives of a function of several variables and may be called the *partial factors* in favour of $H_1, H_2, \ldots, H_n$. Let $P(H_r \mid H) = p_r$. Then *the factor in favour of H in virtue of E is equal to the "weighted average" of the partial factors*, i.e. it is equal to $\sum_r p_r f_r$.

The proof of this is simple. We have

$$\sum_r p_r f_r = \sum \frac{P(H_r \mid H) P(E \mid H_r)}{P(E \mid \bar{H})}$$
$$= \sum \frac{P(H_r \mid H) P(E \mid H_r . H)}{P(E \mid \bar{H})}$$
$$= \sum \frac{P(E . H_r \mid H)}{P(E \mid \bar{H})} = \frac{P(E . H \mid H)}{P(E \mid \bar{H})} = \frac{P(E \mid H)}{P(E \mid \bar{H})}.$$

COROLLARY. The factor in favour of H lies between $\min_r f_r$ and $\max_r f_r$.

Example (i). Imagine an experiment in ESP of the type discussed in 4.9.†
Suppose that there are n trials of which r are successful. Let H denote the

† Section 4.9 should be re-read at this point.

WEIGHING EVIDENCE

hypothesis that the "percipient" has powers of extra-sensory perception. This hypothesis was called K in 4.9. Corresponding to K_κ of 4.9, let H_p be the assertion that the probability is p that a given trial will be successful, and that this probability is a *chance*. Worded in this way, H_p is an "improper theory". The question of whether it could be converted into a proper theory will not be reopened here.

The hypotheses H_p for different values of p are mutually exclusive. If it is assumed that the amount of ESP remains constant, then H is the disjunction of the continuous infinity of propositions H_p for values of p satisfying $\frac{1}{2} < p \leqslant 1$. Let us assume that if H is given then there is a uniform distribution of probability for the variable p between $\frac{1}{2}$ and 1. (See 5.3.) Suppose further that
$$10^{-20} < O(H) < 10^{-3}.$$
What then are the final odds of H in virtue of the whole experiment E?

The "partial factor" in favour of H_p from each success is $2p$ and from each failure is $2(1-p)$. (The factor from a failure is of course less than one.) Hence the partial factor from the whole experiment is $(2p)^r \{2(1-p)\}^{n-r}$. Therefore by the theorem of the weighted average of partial factors,† the factor for H is

$$\int_{\frac{1}{2}}^{1} (2p)^r \{2(1-p)\}^{n-r} 2dp.$$

This could be evaluated by means of tables of the incomplete Beta function. Or we may put $p = \frac{1}{2}(1 + x)$, and, if $\dfrac{r}{n} - \dfrac{1}{2}$ is small, obtain

$$\int_0^1 (1+x)^r (1-x)^{n-r} dx = \int_0^1 (1-x^2)^{\frac{1}{2}n} \left(\frac{1+x}{1-x}\right)^{r-\frac{1}{2}n} dx$$

$$\doteqdot \int_0^\infty \exp\{-\tfrac{1}{2}nx^2 + (2r-n)x\} dx$$

$$= \frac{1}{\sqrt{n}} e^{\frac{1}{2}s^2} \int_{-s}^\infty e^{-\frac{1}{2}y^2} dy,$$

where $s = (r - \tfrac{1}{2}n)/\tfrac{1}{2}\sqrt{n}$. This is the deviation above the mean, divided by the standard deviation, assuming $\bar{H}$. It may be called the "σ-age" of the experiment. If it is at all large (say $s > 2$), while $\dfrac{r}{n} - \tfrac{1}{2}$ is small, a good approximation for the factor is $\sqrt{\dfrac{2\pi}{n}} . e^{\frac{1}{2}s^2}$, a plausibility gain of $(2.17s^2 + 4 - 5\log_{10} n)$ db.

The final plausibility therefore lies between $(2.17s^2 - 196 - 5\log_{10} n)$ db and

---

† This theorem concerns only a finite number of alternatives, but it is adequate for our purpose. For we could work with a large but finite number of alternatives, as in 4.9. The summations to which this would give rise would be approximated by the integrals used here.

69

$(2.17s^2 - 26 - 5\log_{10} n)$ db. For example, if $n = 10,000$ a σ-age of 10 would be required ($r \geqslant 5500$) in order that H should be at least evens.

Many statisticians would be satisfied with a smaller score than this on the grounds that a σ-age of 5 or more is so very improbable on the assumption of no ESP. What this means in effect is that they would take the initial odds $O(H)$ as at least 10^{-4}. This is an application of the " device of imaginary results ", described in 4.3 (iii).

In practice, however, if the number of successes in the first 10,000 experiments really were 5250 it would be suggestive that the assumptions were wrong. It might mean that there was something wrong with the design of the experiment, or that the powers of the percipient were variable. The second hypothesis could be tested by means of the χ^2 test, which will be described in the next chapter. The test could be applied by breaking up the experiment into equal blocks, e.g. 100 blocks each consisting of 100 successive trials, and then seeing if the numbers of successes in the blocks were significantly variable. If no significant variation could be detected and if no fault could be found with the design of the experiment, then the obvious course would be to extend the series of trials. For if the experiment had been worth starting when the probability of success was very low it would presumably be worth continuing when this probability had increased.† The natural time to stop the series of trials would be when the probability had become close to 1, or else appreciably less than it was before the first trial.

Example (ii). The following figures were given as an example in a paper on inverse probability by Haldane (1931).

A family of 400 *Primula sinensis* seedlings from the cross between a doubly heterozygous plant and a double recessive contains 160 " cross-overs ". Let H be the hypothesis that the genes of the original plant lie in the same chromosome. The initial odds of H are 11 to 1 against. Call a cross-over a " failure ", so that there are 240 successes out of 400 "trials". If $\bar{H}$ is assumed the probability of a success is $\frac{1}{2}$, and assuming H, the probability has (approximately) a uniform prior distribution between $\frac{1}{2}$ and 1. What are the final odds of H?

It will be seen that the problem is mathematically identical with the one about ESP which has been discussed above. Here $n = 400, r = 240, \sigma = \frac{1}{2}\sqrt{n} = 10$, $s = 40/\sigma = 4$, so the plausibility gain is $80 \log_{10} e + 4 - 5 \log_{10} 400 = 25 \cdot 7$ db. The initial plausibility is $- 10 \log_{10} 11 = - 10 \cdot 4$ db, so the final plausibility

---

† This argument can be used more generally. It provides some justification for the view that the *factor* from an experiment is of immediate importance, without the direct consideration of the probability of the hypothesis that is being tested. This is true when the decision involved is whether to extend the experiment. It is not true in general for other types of decisions.

is 15·3 db. The final odds are therefore 34 to 1 on, agreeing with Haldane's figure of 0·028 for the final probability of $\bar{H}$.

6.6 Relative factors and relative probabilities

Let $H_1, H_2, \ldots, H_n$ be a set of mutually exclusive and exhaustive hypotheses with probabilities $p_1, p_2, \ldots, p_n$. Any set of numbers proportional to these probabilities may be called the *relative probabilities* of the hypotheses. If E is the result of an experiment, we know that

$$\frac{P(H_r \mid E)}{P(H_r)} \propto P(E \mid H_r).$$

Any set of numbers proportional to the likelihoods $P(E \mid H_r)$ may be called the *relative likelihoods*. With the obvious definition of *relative factors* it is a truism that *the relative final probabilities may be obtained by multiplying the relative initial probabilities by the relative factors*. Moreover the relative factors are equal to the relative likelihoods, by the above form of Bayes' theorem, and therefore, just as in **6.1**, we shall regard the relative likelihoods as providing an alternative definition of the relative factors. If this is done the above " truism " becomes an important theorem.†

Relative factors have a multiplicative property corresponding to T23, when several experiments are performed, provided that these experiments are independent whichever of the hypotheses H_r is assumed.

When there are only two hypotheses H and $\bar{H}$, the ordinary factor is equal to the ratio of the two relative factors, in view of T22. If there are two hypotheses, one of which is composite, the partial factors may be taken as a set of relative factors.

Any sets of numbers of the forms $a + \log P(H_r)$, $b + \log P(H_r \mid E)$, $c + \log P(E \mid H_r)$, where a, b, c are independent of r, may be called the relative initial plausibilities, the relative final plausibilities and the relative weights of evidence. The unit is the bel, the decibel or the natural bel, according as the base of the logarithms is 10, $\sqrt[10]{10}$ or e. Bayes' theorem may be expressed in the form

Relative final plausibilities = relative initial plausibilities
+ relative weights of evidence.

If there are only two hypotheses the theorem reduces to

Final plausibility = initial plausibility + weight of evidence.

This becomes clear when it is observed that if H is a hypothesis, the initial plausibility of H is equal to the relative initial plausibility of H minus the relative

---

† I have now been informed by Dr. C. A. B. Smith that an almost identical formulation of Bayes' theorem is frequently used in population genetics.

initial plausibility of $\bar{H}$, with a similar equality for final plausibilities and weights of evidence.

The notion of relative factors, etc. will be used in the next chapter.

6.7 Expected weight of evidence

There is a curious theorem which was pointed out by Dr. Turing, namely that *the expected factor for a wrong hypothesis in virtue of any experiment is equal to* 1. For example, if an unbiased coin is spun once there is a probability $\frac{1}{2}$ of a factor 0 and also a probability $\frac{1}{2}$ of a factor 2 in favour of the wrong hypothesis that the coin is double-headed. More generally, let the hypothesis be H and suppose that an experiment is performed which must have one of the mutually exclusive results $E_1, E_2, \ldots, E_n$. Imagine that A and B are two people with the same " body of beliefs " but only A knows that H is false. (Assume further that A accepts the theory of probability and that he knows that B does also.) From A's point of view, the expected factor † which B will obtain from the experiment is, by the definition of expectation,

$$\sum_r P(E_r \mid \bar{H}) \cdot \frac{P(E_r \mid H)}{P(E_r \mid \bar{H})} = \sum_r P(E_r \mid H)$$
$$= P(E_1 \vee E_2 \vee \ldots \vee E_n \mid H) = 1.$$

Another slightly paradoxical possibility is provided by the example about dice in **6.1**. Suppose that the hypothesis H is that an unloaded die has been selected, and suppose that, unknown to the experimenter, H is false. The die is thrown once. Then, from the point of view of someone who knows that H is false, there is a probability $\frac{2}{3}$ that the experimenter's degree of belief in H will increase. In other words it is 2 to 1 on that a wrong hypothesis will have its probability increased, in this example.

If, however, the die is thrown an infinite number of times, the experimenter's degree of belief in H will almost certainly tend to 0. In fact, on each throw the expected weight of evidence is much more to the point than the expected factor, because of the additive property of weights of evidence. This property enables T21 of **5.2** to be applied to weights of evidence in a significant manner. The same would not be true for expected factors, since the sum of a number of factors has no particular meaning. It is not surprising that *the expected weight of evidence for right hypotheses is positive and for wrong hypotheses is negative.* This result may be proved with the help of the following inequality, ‡ by taking $p_r = P(E_r \mid \bar{H})$, $p_r f_r = P(E_r \mid H)$. Suppose $p_r > 0, f_r > 0, \Sigma p_r = 1, \Sigma p_r f_r = 1$. Then $\Sigma p_r \log f_r \leq 0$, $\Sigma p_r f_r \log f_r \geq 0$. Equality occurs only if $f_r = 1$ for all r.

† The reader may suspect that this involves the probability of a proposition that is itself concerned with probability. This would contravene the definition of a proposition. But it is clear from the proof of the theorem that the suspicion is ill-founded.
‡ Hardy, Littlewood and Pólya, *Inequalities* (Cambridge, 1934), theorem 9.

In a sequential test of a statistical hypothesis H, it is interesting to know the expected number of trials required for a given gain of plausibility if H is true (or for a given loss of plausibility if H is false). The calculation may be made to depend on the expected plausibility gain from one trial. In fact, a good enough approximation for most practical purposes can be obtained by dividing the required gain of plausibility by the expected gain per trial.

In order to obtain a more precise result, including the distribution functions for the size of the sample, it may be observed that the problem is mathematically the same as a problem of " players' ruin ". Two players, who may be identified with the acceptance or rejection of the lot, play a series of games in which the stakes are equal to the plausibility gain and loss due to a success or failure of the test. Their fortunes are equal to the required gain and loss of plausibility and their probabilities of winning any game are $P(E \mid H)$ and $P(\bar{E} \mid H)$ if H is true, or $P(E \mid \bar{H})$ and $P(\bar{E} \mid \bar{H})$ if H is false. The problem is to find the probability of either players being ruined in a given number of games. This problem is treated by Uspensky (1937), 143. See also Bartlett (1946), where further references may be found.

6.8 Exercises

(i) Show that if the odds against three independent events are o_1, o_2, o_3, then the odds against all three events happening are $(o_1 + 1)(o_2 + 1)(o_3 + 1) - 1$.

(ii) A pack contains an unknown number N of cards each with a different picture on it. A random sample of r cards is taken *with replacement*, and is found to contain s different pictures. Show that the N which receives from this result the maximum relative factor (i.e. the " maximum likelihood " value of N) is the largest N for which

$$-\log\left(1 - \frac{s}{N}\right) > -r \log\left(1 - \frac{1}{N}\right).$$

(iii) With the conditions at the end of **6.7**, if the factors f_r are all close to 1, show that the expected gain of plausibility for H assuming that it is true, is roughly equal to the expected loss of plausibility assuming that it is false. (I am indebted to Dr. Turing for this result.)

(iv) Show that, from the point of view of an experimenter who does not know whether a hypothesis H is true or false, the expected final probability after any experiment is equal to the initial probability. In the same circumstances it is not true in general that the expected final odds are equal to the initial odds.

(v) Let H be the statistical hypothesis concerning a random variable that it is normally distributed with zero mean and unit variance. The only alternative hypothesis is that the distribution is uniform in the interval $(-a, a)$ and vanishes outside this interval. It is decided to take n independent readings and to accept H if it does not lose more than k natural bels, where k may be positive or negative. Show that, from the point of view of someone who knows that H is false, the probability that the experimenter will incorrectly accept H does not exceed

$$\frac{(\pi K)^{\frac{1}{2}n}}{(2a)^n \Gamma(\frac{1}{2}n + 1)},$$

where $K = 2k + n \log \dfrac{2a^2}{\pi}$ and is assumed to be positive. (Dirichlet's integral may be used. See Appendix II.)

(vi) Let H mean that a particular man, known to belong to blood-group A, has a (recessive) gene for blood-group O. Assume that $P(H) = \frac{1}{4}$. His wife belongs to group O and an experiment E consists in testing the blood of their six children and finding that they are all of group A. Assuming that $P(E \mid H) = 2^{-6}$, $P(E \mid \bar{H}) = 1$, prove that $P(H \mid E) = \frac{1}{193}$. (This can be proved mentally in a few seconds.) There is some reason to believe the hypothesis G that the father of the seventh child belongs to group O. It turns out that this child belongs to group O, a result which would be certain given G and would have probability $\frac{1}{386}$ given $\bar{G}$. This provides a factor of 386 in favour of G.

(vii) If in exercise (iii) there are only two possible experimental results, E and $\bar{E}$, show that the expected gain of plausibility if H is true is equal to the expected loss if H is false, provided that $P(E \mid H) = P(\bar{E} \mid \bar{H})$. (It can be proved that the expected gain exceeds the expected loss only if $P(E \mid H) < P(\bar{E} \mid \bar{H})$.)

(viii) Let f be the factor in favour of H from an experiment. Show that the expected value of f^n given H equals the expected value of f^{n+1} given $\bar{H}$. Show also that if H is given, the probability does not exceed g that f does not exceed g.

6.9 Entropy

While the manuscript was with the publishers an article appeared † involving ideas that are related in some ways to those of the present chapter.

Suppose that an event occurs whose probability on *known* evidence is p. It is desired to introduce a simple numerical definition for the amount of information that is thereby conveyed. We have already defined a measure for the weight of evidence in favour of a particular hypothesis, but we are now concerned with the amount of information as such, i.e. the amount from the point of view of a person who is interested merely in collecting information, without reference to any uncertain hypothesis. It is natural to make two

† Shannon, C. E., " A mathematical theory of communication ", *Bell system technical journal*, **27** (July 1948), 379–423.

demands on the measure : (i) it should be a decreasing function of p, and (ii) the amount of information provided by two independent events should be the sum of the separate amounts. The only functions satisfying these conditions are of the form $-\log p$, where the units are natural bels if the base of the logarithms is e. If the base is 2 then the unit may be called an "octave", a "binary digit" or (after J. Tukey) a "bit". For example, if a coin is spun and comes down heads then one bit of information is provided.

Now consider an experiment whose possible outcome is one of a finite (or enumerable) number of mutually exclusive events of probabilities $p_1, p_2, \ldots$ Then the expected amount of information from the experiment is

$$-\Sigma p_i \log p_i.$$

This is called by Shannon the *entropy* of the experiment, by analogy with entropy as defined in statistical mechanics. (See, for example, J. C. Slater, *Introduction to chemical physics* (New York, 1939), 33.)

For a discussion of the properties of the entropy of an experiment the reader is referred to Shannon's article. We content ourselves now with seven simple remarks :—(i) Entropy as defined by Shannon is dimensionless, and the analogous entity in statistical mechanics is, strictly speaking, ordinary entropy divided by Boltzmann's constant. (ii) Shannon refers to the entropy of an "event", but what he calls an "event" is what we call an "experiment". (iii) The distinction between an "experiment" and an "event" has made it possible to introduce entropy in a rather more direct manner than that used by Shannon. (iv) The same units can be used for measuring weights of evidence and entropy. (v) Norbert Wiener has pointed out in conversation that the two sorts of entropy can be identified by introducing a "Maxwell demon". (See Slater, *l.c.*, 45.) (vi) As previously implied, Shannon is not concerned with amounts of information relative to alternative hypotheses. But if we consider such amounts of information we find that, apart from sign, they form a set of relative weights of evidence, in the terminology of page 71. (vii) The weight of evidence in favour of a hypothesis H is equal to the amount of information assuming $\bar{H}$ minus the amount assuming H. Hence the expected weight of evidence is equal to the difference of the entropies assuming $\bar{H}$ and H respectively.

CHAPTER 7

STATISTICS AND PROBABILITY

"... the record of a month's roulette playing at Monte Carlo can afford us material for discussing the foundations of knowledge." KARL PEARSON

7.1 Introduction

Any practical statistical enquiry is concerned with *the numbers of objects of a specified set* ("*individuals*" *of a specified* "*population*") *having various attributes*. The general methods of analysis of the numerical information make up the subject of *theoretical statistics*. This subject can be divided into a " descriptive " part and a " predictive " part. The first part is concerned with such methods of characterising a sample as curve-fitting and the calculation of means and higher moments. In predictive statistics forecasts are made of the properties of a population, given a description of a sample. It is this part of the subject that will be discussed in the present chapter. (Some examples have already occurred in previous chapters.) There is no question here of a comprehensive treatment †—our object is merely to indicate by examples that predictive statistics may be regarded as a branch of probability theory. If it could not be so regarded, probability would have failed to cope with an important class of problems concerning degrees of belief.

Even if predictive statistics is a branch of the theory of probability ‡ it is still often necessary to use somewhat arbitrary procedures in practical work. For sometimes the calculations involved in an exact treatment of a problem are prohibitive. This type of difficulty occurs frequently in other branches of science. For example, it is thought that quantum theory is adequate to explain quite complicated chemical reactions, if only the mathematical equations could be solved. Meanwhile chemists often use other less fundamental theories for their predictions. The difficulty occurs even in pure mathematics. In several good books on mathematical analysis there are topics that are not properly referred back to the axioms. It is believed that rigour is possible but difficult, and a provisional semi-intuitive discussion is felt to be adequate. What is forgivable in pure mathematics is presumably forgivable in the theory of probability.

Many statisticians deny that it is possible to reduce statistics to probability. Their reason is usually connected with the rejection of Bayes' theorem. For example, Fisher considers that his famous principle § of accepting a hypothesis

---
† See the excellent treatises of Cramér, 1946, Kendall, 1945–6, and Wilks, 1944.
‡ It must be emphasised that we are continuing to use the phrase " theory of probability " to mean the theory adopted in this book.
§ Considered by earlier writers, but not systematically.

with maximum likelihood is not deducible from the theory of probability. Neyman and E. S. Pearson, while avoiding the use of Bayes' theorem, have attempted to base statistics on probability by means of " errors of the first and second kinds " and " confidence intervals ". (See **6.2**, **7.4** and **7.10**.) These methods of avoiding the use of initial distributions are valuable, but some subjective judgment is normally required in practice. It is noteworthy that E. S. Pearson † (1947) says in connexion with the 2×2 contingency table : ". . . in a problem of such apparent simplicity, starting from different premises, it is possible to reach what may be very different numerical probability figures by which to judge significance ". He refers also to the " qualities of sound judgment which are the characteristics of a well trained scientific mind ".‡ For us the " different premises " correspond to the different ways in which the contingency table could arise and to the different possible bodies of belief. (Contingency tables will be discussed in **7.9**.)

An attempt has been made to justify a number of statistical procedures by considering their asymptotic properties for large samples. The obvious disadvantage of the use of Bayes' theorem, that the initial probabilities may be " known " only to lie in wide intervals, is likewise overcome by the use of large samples; for large samples produce narrow intervals for the final probabilities. Therefore it seems that any *theoretical* justification of statistical rules should if possible be based on the assumption of small samples. Otherwise it is not convincing that these tests are better than the methods adopted here. The question of a *practical* justification of the use of arbitrary procedures is entirely another matter. It is a question of whether a technique that is theoretically less satisfactory can be practically more convenient. Here the guiding principle is the guiding principle of all science—to use enough common sense to know when ordinary common sense does not apply. The sort of judgment that can be made by common sense is that there are occasions when it is better to be lazy. (Cf. **4.3** (iv).) Such a judgment must be made whenever the chi-squared test or confidence intervals are used. (See **7.8A**, **7.8B** and **7.10**.) The judgments can be expressed in terms of the expected utilities associated with the use of various methods, allowance being made for the gain of time in ignoring some of the information.

7.2 Sampling of a single attribute

The simplest collection of statistics consists of a sample of n objects each of which either has or has not some attribute. Suppose that m of the objects have the attribute. The ratio m/n may be called the *sample frequency (ratio)* or

† See also Barnard, 1947.
‡ The necessity for judgment has never been denied by good statisticians, but it has not often been explicitly emphasised. (But see, for example, Bartlett, 1933, p. 534.)

proportion of the attribute. We shall discuss the connexion between sample frequencies and probabilities. The general conclusion will be that the sample frequency is approximately equal to the probability of the attribute in most cases when n is large. This conclusion is suggested by Borel's theorem. The case $n = 1$ shows that it would be irrational to expect the sample frequency to be exactly equal to the probability.

It is advisable to subdivide the problem according to the type of the sample.

(i) Suppose first that the sample consists of the whole population. In this case there is no need to introduce probabilities into the discussion at all. There are n objects of which m have the attribute, and that is all that needs to be said. But the sample frequency in this case is equal to the probability of the attribute for objects selected at random from the population.

(ii) At the other extreme there are cases when you " know " the probability p before the sample is taken. These cases arise for example in games of chance. More usually you have some initial knowledge, but not sufficient to disregard the value of m entirely. Even in games of chance, if m differed from pn by very much, you would naturally suspect that you had made a mistake in your original judgments. The mistake would usually be that of assuming that some empirical proposition was almost impossible, or that the " trials " were precisely independent.

When the probability p is known before the sample is taken and is unaffected by the results of the sampling, it is a " chance " in the notation of 4.9. In this case the sample can be conceived as having been drawn from a large or even infinite hypothetical population. The chance is sometimes called the " (limiting) frequency in the hypothetical (infinite) population ". This phrase has the advantage of helping some people to gain an intuitive grasp of such problems.

The idea of a hypothetical infinite population can be used quite generally as a method of avoiding talking about " chances ". For example, in the remarks concerning quantum theory in 4.9, $P(E\,|\,H.U)$ can be called " the chance of E given H and assuming that quantum theory is true " or " the limiting frequency of occurrences of E given H in a hypothetical infinite population of trials, assuming that quantum theory is true ". The second description of $P(E\,|\,H.U)$ is sufficiently justified by the " fundamental theorem of probability ". (See 4.10.)

(iii) Next suppose that there is a finite population consisting of a known number N of members, M of which have the given attribute. The number M is unknown, but it is assumed to have an initial probability distribution. You take a random sample with replacement † consisting of n members, of which m

† If N is large it does not make much difference to the numerical results whether the sample is with or without replacement. It is assumed to be with replacement because this case is slightly simpler mathematically.

STATISTICS AND PROBABILITY   7.2

are found to have the attribute. What then is the final probability distribution of M ? And what is the probability that the next member selected will have the attribute ? The second question can be reduced to the first one in the following way :—

Observe first that if the value of M was known then the probability of "success" at the next "trial" would be M/N. Moreover this probability would be a chance † in the sense that it would not be affected by the results of sampling. Now suppose that at any stage the probabilities of $M = 0$, $M = 1$, ..., $M = N$ are assumed to be $p_0, p_1 \ldots, p_N$. These numbers define the probability distribution of the chance. The probability of success at the next trial is

$$p_1 \cdot \frac{1}{N} + p_2 \cdot \frac{2}{N} + \ldots + p_N \cdot \frac{N}{N},$$

by axioms A2 and A3. Hence

T24 *When sampling with replacement, the probability of success at the next trial (given evidence E) is equal to the mean value of the chance of success, the mean value being calculated by using the probability distribution (given E) of the chance.*

A similar result applies for the probability of μ successes in the next ν trials, and may be proved in a similar way.

We return now to the first question. Let H_M denote the hypothesis that M has a particular value also denoted by M. Let $p_0, p_1, \ldots, p_N$ be the initial probabilities of $H_0, H_1, \ldots, H_N$. In virtue of each success the various hypotheses receive relative factors of M/N, and in virtue of each failure they receive relative factors of $1 - M/N$. Hence the relative final probabilities are

$$p_M \left(\frac{M}{N}\right)^m \left(1 - \frac{M}{N}\right)^{n-m}.$$

To obtain the "absolute" final probabilities we must divide the relative final probabilities by their sum. It follows from T24 that the probability of success at the next trial is

$$\frac{\sum_{M=0}^{N} p_M \left(\frac{M}{N}\right)^{m+1} \left(1 - \frac{M}{N}\right)^{n-m}}{\sum_{M=0}^{N} p_M \left(\frac{M}{N}\right)^m \left(1 - \frac{M}{N}\right)^{n-m}}.$$

If N is large it is mathematically convenient to imagine that it is infinite and to replace the chance M/N by a continuous variable x. The point function p_M may then be replaced by a density function $p(x)$ that determines the initial

† It is a population " frequency (ratio) ". In the idealised case of an infinite population it would be a " limiting frequency ".

probability distribution of the chance x. (More generally one could use a distribution function that is not necessarily differentiable.) In terms of $p(x)$ the probability of success at the next trial is equal to

$$\frac{\int_0^1 p(x) x^{m+1} (1-x)^{n-m} dx}{\int_0^1 p(x) x^m (1-x)^{n-m} dx}.$$

For example, if the distribution is uniform, so that $p(x) = 1$, the probability of success at the next trial reduces to $\frac{m+1}{n+2}$. This is sometimes called *Laplace's law of succession*. (The cases $n = 0$, $n = 1$, and $m = n$ are particularly interesting.) It may be deduced that if $m = n$, there is a probability of $\frac{1}{2}$ that the next $n + 1$ trials will all be successful. For by A3 the probability is

$$\frac{n+1}{n+2} \cdot \frac{n+2}{n+3} \cdots \frac{2n+1}{2n+2}.$$

In general, if n is large the function $x^m(1-x)^{n-m}$ has a very sharp peak at $x = m/n$. It follows that the probability of success at the next trial is close to m/n, provided that the graph of $p(x)$ has a moderate area in the neighbourhood of $x = m/n$. In other words, if n is large the result is not sensitive with respect to the assumed initial probability distribution of the chance. This is just as well because it is often artificial to give the initial distribution at all exactly.

(iv) Now suppose that the population is infinite. This case cannot really occur except as an idealisation, and in this sense it has already been discussed under heading (iii). It might be thought that infinite populations do occur in such experiments as dice-throwing, but even here the dice would eventually get worn out. It is necessary to fix the value of N in any such case in order to bring it under heading (iii), but the value selected makes very little difference provided that it is large. There is here no question of sampling with replacement, so the previous discussion requires some modification. But the modification presents no particular difficulty and will not be given here.

Instead of regarding this case as being included under heading (iii) it may be more convenient to make direct judgments about the initial distribution of the chance. For example, if this distribution is uniform the sample frequency, m/n, is the "most probable value" † of the chance. (Whatever the initial distribution the sample frequency is the maximum likelihood value of the chance.)

If n were large, adherents of the frequency approach would *judge* that the chance x was approximately m/n. (They would not usually judge that the proportional accuracy was good if m was small.) If they would define the degree

† See the index.

STATISTICS AND PROBABILITY 7.3

of the approximation then Bayes' theorem (in reverse) could be used for obtaining information about the initial probability distribution of the chance.

(v) Finally, suppose that N is unknown. As before you can use judgments about the initial distribution of the chance. (Or you could work with the distribution of N and the distribution of M for each N.)

7.3 Example

Consider the ESP experiment of 4.9 and 6.5. Here the alternative hypotheses are $H_p (\frac{1}{2} < p \leqslant 1)$, where $H_{\frac{1}{2}}$ is the same as $\bar{H}$. Let the initial odds of H be 10^{-10}. If there are m successes in n trials and if it is assumed that there is a uniform initial distribution for p in the range $\frac{1}{2} < p \leqslant 1$, then the relative final probabilities of the alternatives are

$$P_{\text{rel}}(\bar{H} \mid E) = 10^{10},$$
$$P_{\text{rel}}(dp \mid E) = (2p)^m \{2(1-p)\}^{n-m} dp \qquad (\tfrac{1}{2} < p < 1),$$
$$P_{\text{rel}}(p = 1 \mid E) = 0,$$

where some self-explanatory notations have been used. The last of these equations may be denied on intuitive grounds, but it follows from the assumption of a uniform initial distribution. It may be more natural to allow a very small probability to the hypothesis that the man has perfect ESP, but it would not introduce any new interest or difficulty into the calculations. It follows from 6.5 that the final probability of H is large if $2 \cdot 17 s^2 - 5 \log_{10} n - 96$ is large, where $s = (m - \tfrac{1}{2}n)/(\tfrac{1}{2}\sqrt{n})$. Under the same circumstances it is fairly clear that the probability of success on the next trial will be close to m/n. If, on the other hand, $2 \cdot 17 s^2 - 5 \log_{10} n - 96$ is negative and numerically large, then $\bar{H}$ will remain highly probable and the probability of success at the next trial will be very close to $\frac{1}{2}$. In any case, provided that n is large, the probability of success at the next trial is close † to m/n. This is an example of the fundamental theorem of probability.

It should be noticed that if n is not large enough, then the probability of success at the next attempt may be quite different from m/n. For example, if $m = n = 20$, the probability is still close to $\frac{1}{2}$, assuming that $O(H) = 10^{-10}$.

This ESP experiment exemplifies the important ideas of *significance* and *estimation*. If m is sufficiently far from $\frac{1}{2}n$ then ESP is probable and the experiment is called significant. In this case it becomes interesting to know how much ESP is present—that is, which of the hypotheses H_p is true, where $p > \frac{1}{2}$.

The example is typical of many others, and it frequently happens, at any rate as a sufficiently good approximation, that there is a finite amount of the

---

† The reader may consider what modifications are required to allow for the possibility that m is much *smaller* than $\frac{1}{2}n$.

initial probability concentrated at a particular value of a parameter, all other values of the parameter being almost impossible. But in most cases the probability that the parameter has the special value is not so near to 1. For example, if you were investigating whether cosmic rays have any influence on mutation rates of *drosophila* (flies), the initial probability could reasonably be taken as lying between 0·01 and 0·99. There is no need for the parameter to represent a chance. It might for example be a function (such as the mean value) of the chance distribution of the increase in weight of guinea-pigs when injected with a particular drug.

7.4 Inverse probability versus " precision "

Let us say that one probability is *more precise* than another one if it is known or judged to lie in a narrower interval, and that a probability is *precise* if the interval reduces to a point. (See 4.3 (i).) Most tautological probabilities are precise.

Let E be the result of an experiment † (e.g. "heads" or "tails"). If H is a hypothesis it sometimes happens that $P(E \mid H)$ is precise whereas $P(H \mid E)$ and $P(H)$ may not be. Suppose further that the experiment is merely one of a sequence of similar experiments (or trials) and that the probability of E, given H, is a chance in the sense that it is unaffected by a knowledge of the results of other experiments of the sequence. Then H is called a *simple statistical hypothesis*. The whole sequence of trials may be regarded as a sample from an infinite population, in which $P(E \mid H)$ is the limiting frequency of results of a particular "kind". (In die-throwing there are six "kinds" of results.)

If H is a disjunction of a set of mutually exclusive simple statistical hypotheses, then H is called a *composite statistical hypothesis*.‡ In **6.5**, for example, H_p is simple for each p and H is composite. Another example of a simple statistical hypothesis is the assertion that a chance distribution is normal with zero mean and unit variance. This would have been composite if the mean and variance had not been specified.

With this terminology, the likelihood of a simple statistical hypothesis is precise, although its initial and final probabilities may not be. The absolute precision of the likelihoods is usually purchased at the expense of expressing the hypothesis in the form of an incompletely defined proposition.

Given a set of statistical hypotheses, Fisher's principle of maximum likelihood tells you to select that hypothesis whose likelihood is greatest. If the result is unique the procedure is a precise one and does not depend on a subjective judgment of the initial probabilities of the hypotheses. (Cf. **6.8**,

---

† Or rather the proposition asserting what this result is. (See **4.2**.)
‡ These definitions are a little more general than those usually given. See, for example, E. S. Pearson, 1942, 311.

exercise (ii) and **7.2** (iv).) The principle of maximum likelihood is not the only precise procedure that is possible. Another (trivial) one is that all hypotheses should be rejected.

If the hypotheses depend on a single parameter the " maximum likelihood value of the parameter " is equal to the most probable value, provided that the parameter has a uniform initial distribution. If the maximum in the final distribution is " sharp ", then the parameter has a high probability of being close to the maximum likelihood value. Cases approximating to this are fairly common, so that the practice of using maximum likelihood values can often be justified in terms of the theory of probability.

Precise procedures are convenient and often time-saving. But a man's decisions are normally based on what he really believes, i.e. on the final probabilities of the hypotheses. In economics and sociology the samples are usually large and the final probabilities are insensitive to the initial ones. But in many biological experiments the samples are small and then the initial probabilities should be taken into account. These experiments are usually designed to test a plausible hypothesis. If the initial probability is judged to be as high as 0·05, then a factor of 20 would be sufficient to make the hypothesis " odds on ". But different biologists may naturally have different opinions about the initial and therefore the final odds. One objective in using precise procedures is to avoid these differences of opinion. We may be sure that this objective will not be attained. For example, very few scientists would accept a theory based on superstition, even if it received a factor of 1000 from the first experiment. It may be argued that this sort of thing would not happen very often. But *in any given case* what really matters is the final probability of the theory. And besides, it is always possible, when there are far more people engaged on medical and biological research, that it will be quite usual to test hypotheses with very low initial probabilities. A correspondingly larger factor would then be required before a hypothesis would become acceptable. This shows how arbitrary is any rule that depends only on the likelihoods.

Another procedure that may at first sight appear to be precise is afforded by the technique of " errors of the first and second kinds " introduced by Neyman in 1930. (See Neyman (1941) and Neyman and Pearson (1933, twice).) Let E be " an experiment ". Let H and H' be mutually exclusive simple statistical hypotheses. Suppose that no hypothesis other than H and H' needs consideration, i.e. it is judged to be adequate to suppose that $H \vee H'$ is true. Even if there are other plausible hypotheses it is often convenient to deal with only two at a time. It is usually interesting to know the odds of H, but we may have to be satisfied with the ratio of the probabilities of H and H'. By regarding $H \vee H'$ as given, the problem is in any case reduced to the consideration of only two alternative hypotheses. This is convenient because it makes the language of odds and factors more appropriate. The probability of H' is the same as the probability of $\bar{H}$, if $H \vee H'$ is given. We shall use the "misleading notation" of omitting $H \vee H'$ to the right of the vertical stroke. With this understanding $\bar{H}$ can be written instead of H'. (Cf. **6.3**.)

Now suppose that a precise procedure has been described for calculating a "function" $\mathcal{P}(E)$ of the observations, whose possible values are the instructions "reject H" or "accept H". We say that *an error of the first or second kind* (with respect of H) is committed if H is rejected when true or accepted when false, respectively. (Clearly an error of the first kind with respect to H is an error of the second kind with respect to $\bar{H}$, and vice versa.) For the given procedure $\mathcal{P}$, the probability given H of an error of the first kind, and the probability given $\bar{H}$ of an error of the second kind, can be calculated exactly. If it is decided that these probabilities must not exceed two values α and β, a restriction will be provided on the possible procedures $\mathcal{P}$. For example, in exercise (v) of **6.8**, the probability of an error of the second kind (when $\bar{H}$ is given) is less than β if

$$k < \frac{2a^2}{\pi} \left\{ \Gamma(\tfrac{1}{2}n + 1)\beta \right\}^{\frac{2}{n}} - \tfrac{1}{2}n \log \frac{2a^2}{\pi}.$$

When α and β are given, $\mathcal{P}$ is a precise procedure. But the choice of α and β depends on judgment. (Cf. **6.2**.)

Other methods of avoiding the use of the initial probabilities of hypotheses will be discussed in **7.8A**, **7.8B** and **7.9**, in connexion with the chi-squared test, and also in **7.10**.

7.5 Sampling and the probabilities of chance distributions (curve-fitting)

Consider the heights to the nearest inch of a population of men. Suppose for the moment that the heights of all the men in the population are known and that a man is selected at random. The chance of his having any particular height is known. (We call it a chance because it is independent of any sampling.) Thus the chance distribution of the heights is known, rather than the probability distribution. Assuming next that only the size of the population is known and that no man can be more than 20 feet high, then the number of possible chance distributions is finite. Hence you can associate with each distribution a finite probability which will depend on the evidence assumed. The set of such probabilities defines what may be called the *probability distribution of the chance distribution*. This distribution of distributions is known only vaguely before a sample is taken. The question is how much can be said about it afterwards. This is a central type of problem in statistics. (Cf. **5.4**.)

We shall idealise the problem to the extent of assuming that the population

STATISTICS AND PROBABILITY   7.5

is infinite † as in **7.2** (iii) and (iv). This will have the effect of making the chance distribution continuous rather than discrete. It may lead on to a consideration of measure in function space, as mentioned in **5.4**, but in practice the chance distribution is usually judged to be defined adequately by means of only a finite number of parameters.

Since the population is assumed to be infinite it does not matter whether the sample is with or without replacement, but for definiteness it may be assumed to be without replacement.

Each particular number of inches is an attribute. Thus the problems that arise are more complicated than before when there was only one attribute. The previous discussion shows that the probability that the next man selected will have a given height (to the nearest inch) is roughly equal to the sample frequency, provided that the sample is large enough. This shows the similarity with the problem of sampling a single attribute. But there is a new consideration that is roughly expressed by the idea of smoothness. This will be explained by means of an example.

Suppose that the sample consists of 1000 men, the numbers in the various groups being given by the following table :—

Height in inches	59	60	61	62	63	64	65	66	67	68	69	70	71	72	73	74	75
Numbers of men	1	3	12	23	53	73	96	156	150	157	118	83	39	19	12	4	1

TOTAL NUMBER OF MEN : 1000

What is the probability that the next man selected will have a height of 67″? The table, or a graph constructed from it, suggests that the probability is greater than 0·150. You are influenced by a feeling that a graph of the chances ought to be smooth,‡ i.e. that it should not have many " bumps ".§ Thus the probability is affected not only by the number of men of height 67″, but also by all the other entries in the table, and especially by the entries under 66″ and 68″. It may be asked where you get this belief in smoothness, and whether rules can be given for deciding the probabilities more precisely.

These questions can hardly be answered completely since they depend on probability judgments. Perhaps the main point involved is the principle of simplicity. This asserts that a simple hypothesis has a higher initial probability than a complicated one. The question has already been discussed in **5.4**. We referred in **5.4** to the number of parameters involved in the analytic expression of a function, as a measure of its simplicity. Another possible measure is

---

† This device is sometimes used even when a sample consists of the whole of a population. In this case it may be helpful to imagine that the population is itself merely a sample of an infinite " super-population ".
‡ i.e. you associate higher initial probabilities with smooth chance distributions.
§ E. S. Pearson, 1938, defines smoothness in terms of Legendre polynomials.

the *number of points of inflexion*, this being a natural measure of " bumpiness ". Thus in the present example a better fit to the observations could be obtained by means of a " double-humped " curve † (which has four points of inflexion), but a single-humped curve may seem more probable. (It need have only two points of inflexion.) Another reason for preferring the simpler curve is that any given simple curve is found in practice to occur, as an approximation, more often than any given complicated curve.

In particular, single-humped curves occur more often in connexion with cases similar to the one under consideration, provided that the sample is large. More precisely it is known by experience that small bumps tend to get smoothed out when the size of the sample is increased, the class-interval being kept constant. Thus *the statistics of statistics* have some influence on your opinions. (See also the last paragraph of this section.)

Besides the initial probabilities of the chance distributions you need to consider the factors obtained from the sample. Suppose that a particular chance distribution is assumed in which the chance of a height of r inches is p_r. Suppose further that the number of men of height r inches in the sample is m_r, where $\sum_{r=0}^{\infty} m_r = n$. Then the relative factor in favour of this distribution may be taken as

$$\prod_{r=0}^{\infty} p_r^{m_r},$$

where 0^0 is defined as 1. (The multinomial coefficient is omitted since it is the same for all distributions.)

As an example consider the chance distributions that are of the normal form

$$\frac{1}{\sigma\sqrt{2\pi}} e^{-(x-x_0)^2/2\sigma^2}.$$

The chance of a height of r inches is then

$$p_r = \frac{1}{\sigma\sqrt{2\pi}} \int_{r-\frac{1}{2}}^{r+\frac{1}{2}} e^{-(x-x_0)^2/2\sigma^2} dx.$$

Thus the relative weights of evidence may be taken as

$$- n \log \sigma + \sum_{r=0}^{\infty} m_r \log \int_{r-\frac{1}{2}}^{r+\frac{1}{2}} e^{-(x-x_0)^2/2\sigma^2} dx.$$

† The observations could be fitted exactly by means of a polynomial of the 16th degree, but the result would be far too complicated to be regarded as a probable distribution of the chance.

STATISTICS AND PROBABILITY 7.5

The derivative of this with respect to x_0 is

$$\frac{1}{\sigma^2}\sum_{r=0}^{\infty}\left\{m_r\int_{r-\frac{1}{2}}^{r+\frac{1}{2}}e^{-(x-x_0)^2/2\sigma^2}(x-x_0)\,dx \Big/ \int_{r-\frac{1}{2}}^{r+\frac{1}{2}}e^{-(x-x_0)^2/2\sigma^2}\,dx\right\}.$$

The coefficient of m_r is approximately † $(r - x_0)$. It follows that the maximum likelihood value of x_0 is approximately $\frac{1}{n}\Sigma rm_r$, the average height of the men in the sample. In a similar way the maximum likelihood value of σ^2 is approximately ‡ $\frac{1}{n}\Sigma m_r(r-x_0)^2$. For any assumed initial distribution of x_0 and σ the final distribution can be written down. The maximum likelihood values of x_0 and σ will be close to their expected values under natural assumptions concerning the initial distributions, provided that n is not too small. The combined distribution of x_0 and σ defines the distribution of the chance distribution. From this you can calculate the probability that the next man selected will have any particular height, i.e. the final probability distribution of the height of the next man to be selected. This is the sort of thing that would normally be of most interest in such problems.

In order to save work you could assume that this final result is sufficiently well approximated by a normal distribution in which the parameters x_0 and σ are taken as equal to their expected values, or even to their maximum likelihood values. Using the latter method with the given figures, it is found that $x_0 = 67 \cdot 00$, $\sigma = 2 \cdot 536$, and the values of $1000 p_r$ are given in row (iii) of the following table :—

	59	60	61	62	63	64	65	66	67	68	69	70	71	72	73	74	75	Total
(i)	59	60	61	62	63	64	65	66	67	68	69	70	71	72	73	74	75	Total
(ii)	1	3	12	23	53	73	96	156	150	157	118	83	39	19	12	4	1	1000
(iii)	0·3	3·4	10	23	46	79	115	145	156	145	115	79	46	23	10	3·4	0·3	999·4
(iv)	0·3	3·4	10	23	46	75	111	156	150	157	110	75	46	23	10	3·4	0·3	999·4
(v)						−16·4	−14·8	49·4	−25·5	54·2	−22·8	−18·7						5·4

The meanings of the rows of this table are :—(i) Height in inches ; (ii) sample figures ; (iii) maximum likelihood normal curve ; (iv) a double-humped curve ; (v) plausibility gain of (iv) minus that of (iii), in db.

Row (iv) has been selected so as to fit the observations better than row (iii) in the neighbourhood of the mean. The idea is to test the hypothesis that the chance distribution is really double-humped. For this purpose it would be a mistake to make row (iv) agree too well with row (ii) in the "tails". The double-humped curve is only 5·4 db "better" than the normal one, and is therefore hardly to be preferred, after allowing for the relative initial probabilities.

† There is a lack of rigour here. The approximation is not good where $|r - x_0|/\sigma$ is greater than 2, but in our example m_r is small for such values of r.

‡ There are reasons for preferring the estimate $\frac{1}{n-1}\Sigma m_r(r-x_0)^2$ for σ^2. (See, for example, Wilks, 83.)

It is often found that the normal distribution or some other standard distribution is a good fit except in the tails. Such a modified result is simple enough to have an appreciable initial probability and is useful, although it does not enable you to estimate the probabilities of very rare events with much proportional accuracy.

Besides the various standard distributions the possibility of using a linear combination of them should always be borne in mind. This is especially suggested if the population is considered initially to be likely to be composed of two or more types. For example, the heights of all adults in England might be expected to obey roughly a distribution equal to a linear combination of two normal distributions corresponding separately to men and women.

We conclude this section with another remark about smoothness. So far we have attempted to justify the assumption of smoothness in terms of simplicity and past experience. There is another possible justification. It has been found that the convolution of a large number of independent distributions tends to be smooth, even though the original distributions are not. For a treatment of such problems the reader is referred to Jessen and Wintner (1935). Their results justify the assumption of smoothness in the same way that the central limit theorem justifies the normal distribution, but rather more vaguely. It seems likely that there would be similar results even if the distributions that were "compounded" were not entirely independent.

7.6 Further remarks on curve-fitting

In this section we shall refer briefly to some standard methods of curve-fitting. This will be done not in order to explain the methods,† but to indicate roughly their relation to the theory of probability and to the remarks of the previous section.

A system of curves was defined by Karl Pearson, depending on only four parameters, and giving an adequate representation of many single-humped curves as well as some J-shaped and U-shaped ones. The system is used a great deal by statisticians, the usual method of fitting being by means of the first four moments of the "observed distribution" (i.e. the frequency distribution of the sample). The system may be described as a simple one, partly because there are only four parameters and partly because no curve of the system can have more than two points of inflexion. The normal distributions are included together with other classes of curves that arise naturally from a theoretical point of view. For these reasons the system may be considered to have a moderate initial probability of applying approximately in any given case. The system is used also because of its convenience. It often happens that the approximation is not good over the whole range. This is hardly

† See Kendall, 1945, and Elderton, 1938.

STATISTICS AND PROBABILITY 7.7

surprising since smoothness of a curve does not imply simplicity of its analytic expression. Above all, it is the smoothness of a distribution that seems to give it an appreciable initial probability. It is often adequate to draw free-hand curves instead of doing calculations. Another method is to fit parabolas to different parts of the frequency curve; but the initial probability would presumably be taken as a rapidly decreasing function of the number of parabolas.

A theoretically correct method of carrying out this curve-fitting is to make use of the relative factors as in **7.5** and to allow for the initial probabilities of the possible curves. In practice it is necessary to simplify $\mathfrak{B}$, as in **4.3**, suggestion (iv), by making allowance only for a particular system of curves, such as the Pearson system. In this case the initial distribution of the curves is fixed by the initial distribution of the parameters. It may often be judged † that the method of maximum likelihood will give adequate results.

There are other standard methods of curve-fitting besides those already mentioned. One of these is by expansion in Hermite functions; another is by the transformation of the variable so as to obtain approximately a normal distribution of the new variable. Of these two methods the second seems to have more justification from the point of view of the theory of probability. The possibility of obtaining a rough initial justification of the type of curve should not be overlooked.

7.7 The combination of observations

In many physical experiments several measurements are made of what is supposed to be the "same" physical magnitude. It is usually supposed that there is a true value, that the deviations from this are due to the accumulation of unavoidable errors, and that these deviations obey a normal distribution.‡ The assumption that there is a true value may be avoided by assuming merely that the possible results of the experiment obey a normal distribution with mean x_0 and variance σ^2. The parameter x_0 takes the place of the so-called true value. The problem of estimating x_0 and σ is then mathematically as in **7.5**.

Suppose that one of the readings is a long way from the rest of the observations. Is it justifiable to reject it when estimating x_0 and σ? The answer depends on the probability of having made a mistake, i.e. an avoidable error. In general, if the deviation from the average is greater than five times the estimated value of σ, it would probably be assumed that a mistake had been made, because the factor in favour of a mistake would be large. Or the assumption of a normal distribution might be suspected unless it was well supported by the rest of the observations.

† The judgment is one that can be expressed in terms of expected utilities in any particular case. But the implicit courses of action themselves involve alternative probability techniques, so that the judgment is of a higher "type" than usual.

‡ The mean need not be zero, i.e. there may be a bias.

7.8   PROBABILITY AND WEIGHING OF EVIDENCE

The exact values of x_0 and σ cannot be determined. All that can be done is to say something about their probability distributions, maximum likelihood values and so on. For a detailed account of the subject see Brunt (1931).

Exercise. Only two theories are to be entertained regarding the value of a physical magnitude : either it is equal to ξ or else to ξ'. Several experiments are performed and readings $x_1, x_2, x_3, \ldots$ are obtained. Assuming a normal law of error with standard deviation σ, show that the first theory gains

$$\frac{\xi - \xi'}{\sigma^2} \sum_r \left(x_r - \frac{\xi + \xi'}{2} \right) \text{ natural bels.}$$

7.8 Significance tests

The general question of significance tests was raised in **7.3** and a simple example will now be considered. Suppose that a die is thrown n times and that it shows an r-face on m_r occasions ($r = 1, 2, \ldots, 6$). The question is whether the die is loaded. The answer depends on the meaning of "loaded". From one point of view it is unnecessary to look at the statistics since it is obvious that no die could be absolutely symmetrical.† It is possible that a similar remark applies to all experiments—even to the ESP experiment, since there may be no way of designing it so that the probabilities are *exactly* equal to $\frac{1}{2}$. In the case of the die let us suppose that it has chances $p_1, p_2, \ldots, p_6$ of showing a 1, 2, ..., 6, these chances being initially unknown. We could say that the die is loaded if for example

$$\sum_{r=1}^{6} |p_r - \tfrac{1}{6}| > \tfrac{1}{100}.$$

Suppose that there is an initial probability density of the chances, given by a function $\varphi(p_1, p_2, \ldots, p_6)$. This is defined in such a way that if V is any five-dimensional volume of the space $\Sigma p_r = 1$, the probability that $(p_1, p_2, \ldots, p_6)$ belongs to V is equal to $\int_V \varphi \, d\tau$, where $d\tau$ is an element of volume. The function φ depends on your body of beliefs and on your knowledge concerning dice in general and on where the particular die was obtained. It is convenient to take the relative factors in such a way that the factor corresponding to symmetry is 1. Then the relative factor for the set of chances $p_1, p_2, \ldots, p_6$ is $\prod_r (6p_r)^{m_r} = f$ say. The final probability that the die is loaded is

$$\int_{D_1} \varphi . f \, d\tau \bigg/ \int_{D} \varphi . f \, d\tau,$$

where D is the space $\Sigma p_r = 1$ and D_1 is the sub-space in which $\Sigma |p_r - \tfrac{1}{6}| > \tfrac{1}{100}$.

† It would be no contradiction of 4.3 (ii) to say that the hypothesis that the die is absolutely symmetrical is almost impossible. In fact, this hypothesis is an idealised proposition rather than an empirical one.

STATISTICS AND PROBABILITY

For any definite assumption concerning φ, this probability has a numerical value that may be difficult to calculate. In practice you have to be satisfied with approximations.

If you were doing the problem rigorously you would half-define φ by means of inequalities, possibly rather vague, but not depending on n or $m_1, m_2, \ldots, m_6$. In order to obtain an approximate result you could take simplified assumptions for φ. These assumptions *would* depend on the results of the statistics. They would depend also on the properties of the relative factor f. As a function of $p_1, p_2, \ldots, p_6$, f has a maximum at $p_1 = \frac{m_1}{n}$, $p_2 = \frac{m_2}{n}, \ldots, p_6 = \frac{m_6}{n}$. This maximum is fairly sharp, so that the values of φ at points far removed from the maximum have little effect on the values of the integrals. An exception must be made of points where φ is especially large. Regarding the correct form of φ, it would be irrational to assume that it vanished at any point where $\Sigma p_r = 1$. But for an approximation you could regard zero values of φ as admissible, and change your mind if the statistics suggested that you should. In particular, if the sample were not too large, the density function in the sub-space $\Sigma |p_r - \frac{1}{6}| \leqslant \frac{1}{100}$ could be replaced by a point function vanishing at all points except at $p_1 = p_2 = \ldots = p_6 = \frac{1}{6}$. The value of the point function here is the initial probability that the die is unloaded. Call it $1 - p$. Let $q = \int_{D_1} \varphi . f d\tau$. Then $\int_{D} \varphi . f d\tau = q + 1 - p$, and the final probability that the die is loaded is equal to $q/(1 - p + q)$. Therefore the final odds are $\frac{q}{1-p}$ and the factor in favour of the die being loaded is

$$\frac{q}{p} = \int_{D_1} \psi . f d\tau,$$

where $\psi = \frac{\varphi}{p}$ = the initial probability density of the chances given that the die is loaded. The formula $\int_{D_1} \psi . f d\tau$ could also have been deduced directly from the theorem of the weighted average of factors.

One consequence is that the factor in favour of the die being loaded cannot exceed max $f = \prod_r \left(\frac{6m_r}{n}\right)^{m_r}$. If it is assumed that $\frac{6m_r}{n} - 1$ is small, for all six values of r, it follows easily that the weight of evidence does not exceed

$$\tfrac{1}{2} \sum_{r=1}^{6} \frac{\left(m_r - \frac{n}{6}\right)^2}{\frac{1}{6}n} \text{ natural bels.}$$

We write the weight of evidence in this form in order to exhibit its connexion with the chi-squared test. (See **7.8A**.) It may be observed that this result does not depend on ψ. Next we shall work out the factor for a particularly simple assumption concerning ψ. It is hardly necessary to point out that the results would be different with different bodies of beliefs.

In the first place suppose that if the die is loaded then it is loaded in such a way as to make p_6 larger than any of $p_1, p_2, \ldots, p_5$. Assume, in fact, that if the die is loaded then the chance of a 6 is uniformly distributed † between a and b where $b > a > \frac{1}{6}$. Assume further that $p_1 = p_2 = p_3 = p_4 = p_5$, so that each is equal to $\frac{1-p_6}{5}$. With these assumptions $\psi = \frac{1}{b-a}$ where $a < p_6 < b$, and the factor is

$$\frac{1}{b-a} \int_a^b \left(6 \cdot \frac{1-p_6}{5}\right)^{n-m_6} (6p_6)^{m_6} dp_6.$$

If n is small this can be calculated exactly. If n is large and m_6 is not close to $n/6$, then the die is obviously loaded and the calculation is unnecessary. Finally, if n is large and m_6 is not too far from $\frac{n}{6}$ the factor can be calculated by means of the following rough argument.

The integrand has a maximum at $p_6 = m_6/n$. In the neighbourhood of the maximum the logarithm of the integrand is approximately equal to $\frac{n}{10}(\mu^2 - x^2)$, where $\frac{m_6}{n} = \frac{1}{6}(1+\mu)$, $p_6 = \frac{m_6}{n}(1+x)$. (The analysis is straightforward.) It follows that the factor is approximately

$$\frac{e^{n\mu^2/10}}{6(b-a)} \sqrt{\frac{10\pi}{n}}.$$

For example, if $b - a = \frac{1}{3}$ the weight of evidence is

$$\left\{ \frac{3}{5} \cdot \frac{\left(m_6 - \frac{n}{6}\right)^2}{\frac{1}{6}n} + \frac{1}{2} \log\left(\frac{5\pi}{2n}\right) \right\} \text{natural bels,}$$

and this may be compared with the approximate form of the maximum weight of evidence. As an example of the present formula suppose that $n = 600$, $m_6 = 140$ and the initial odds are between 0·001 and 0·01; then the factor is about 2000 and the final odds are between 20 to 1 on and 200 to 1 on that the die is loaded. This assumes that none of the numbers $m_1, m_2, \ldots, m_5$ shows any considerable deviation from 100. If the only large deviation were on, say, m_1 instead of m_6, then the factor would be the same, but the initial odds that the die was loaded in this way would be much smaller. If these odds were between 0·0001 and 0·001 the final odds would be between 5 to 1 against and

† The possibility of p_6 being less than $\frac{1}{6}$ could also be taken into account with only slight modifications.

2 to 1 on. A similar adjustment could be made if m_6 had been far below the mean instead of far above. Finally, if more than one of the numbers m_r showed a large deviation, it would be necessary to sharpen the argument. The weight of evidence would presumably come out as a sum of expressions resembling the one given above.

7.8A The chi-squared test

So much for the solution of the problem based directly on Bayes' theorem. Many statisticians would have used the chi-squared test. The idea of this test is to take a particular function of the statistics, a function that for this particular problem † is

$$\chi^2 = \sum_{r=1}^{6} \frac{(m_r - \tfrac{1}{6}n)^2}{\tfrac{1}{6}n},$$

and to work out the probability that this random variable is greater than or equal to the value actually attained (say χ_0^2), on the hypothesis that the die is symmetrical. Let this probability be denoted by $P(\chi_0^2)$.‡ If it is assumed that the corresponding probability when the die is loaded is close to 1, then $P(\chi_0^2)$ may be regarded as the factor in favour of the die being symmetrical in virtue of the knowledge that $\chi \geqslant \chi_0$. This is not the same as the factor in virtue of the whole experiment, since some of the evidence is ignored. The true factor depends on all the numbers $m_1, m_2, \ldots, m_6$, whereas in the chi-squared test only the value of $\Sigma(m_r - \tfrac{1}{6}n)^2$ is used. Moreover the factor is worked out on the evidence that $\chi \geqslant \chi_0$, but really you know that $\chi = \chi_0$. It might be suggested that the result of the experiment could just as well be expressed as $\chi \leqslant \chi_0$ instead of $\chi \geqslant \chi_0$. But if χ_0 was large so much evidence would be thrown away by this alternative procedure that the resulting factor would be close to 1. (In fact the likelihoods of the hypotheses " loaded " and " unloaded " would both be near 1.)

As already pointed out, you really know that $\chi = \chi_0$. Since χ_0 can be known only to a certain number of places of decimals, the factor worked out by regarding $\chi = \chi_0$ as the result of the experiment is not of the "indeterminate" form 0/0, though the numerator and denominator are both small. As an approximation the distribution functions of χ (or χ^2), on the two hypotheses "loaded" ($\bar{H}$) and "unloaded" (H), could be assumed to have density functions. The factor in favour of H is then the ratio of these density functions at χ_0. The denominator would not usually be known at all precisely. It could be estimated either by a direct judgment or by calculations based on other judgments. It might be assumed, for example, that, given $\bar{H}$, the graph of the

---

† For a more general definition of χ^2 see **7.8B**.
‡ This is admittedly a rather unsatisfactory notation in the present context.

distribution of χ^2 is obtained approximately by averaging for all λ between 0 and say $\frac{1}{8}$ the results obtained by shifting the graph of the χ^2 distribution (given H) through a distance λn to the right.

It would often happen that the factor in favour of H obtained in some such way would be in the region of three or four times $P(\chi_0^2)$.† From the present point of view this is the main justification for using $P(\chi_0^2)$ as a measure of the significance of the experiment. Some statisticians would say that the chi-squared test has nothing to do with Bayes' theorem and that it simply seems rational to estimate significance by calculating the probability of χ being as large as χ_0 or larger. This so-much-or-more idea is very arbitrary and easy to criticise. An alternative justification of the chi-squared test is available by means of the Neyman-Pearson technique of errors of the first and second kind. But, just as in the inverse probability method, this technique is applicable only if something is assumed about the distributions of χ^2 given both the hypothesis being tested and its negation.

A weakness of the chi-squared test, for the problem of the die, is that it does not take into account the peculiar significance of the "6"-face. We should like to be able to give additional weight to the term $(m_6 - \frac{1}{6}n)^2/\frac{1}{6}n$. In more general problems it would be useful to know the distribution of *any* linear form in the numbers analogous to $(m_r - \frac{1}{6}n)^2/\frac{1}{6}n$, instead of only the sum. As far as I know this problem has not been solved in a convenient mathematical form.

In view of the difficulties of a strict application of Bayes' theorem and in view of the criticisms of the chi-squared test, perhaps the best practical procedure is something intermediate. For example, you could use the chi-squared test, and take $1/4P(\chi_0^2)$ as the approximate factor in favour of a hypothesis to be stated *after seeing the statistics*. The initial probability of this theory could then be judged subjectively. For example, if the main deviation were on the 1's and the 3's, you could take as your hypothesis that "the die is loaded but not with respect to 6's" and perhaps judge that the initial probability lies between 0·0001 and 0·001. Here it would not be right to formulate the hypothesis in terms of 1's and 3's (which would decrease the initial probability still more) since in using the chi-squared test no credit is allowed for the fact that the main deviations are with respect to these particular faces.

Another point about the chi-squared test is that if n is very large, the test will probably give a significant result, because the chances, $p_1, p_2, \ldots, p_6$ can

---

† There are two independent reasons why the factor in favour of H exceeds $P(\chi_0^2)$. The first is that to pretend that the result is $\chi > \chi_0$ when it is really $\chi = \chi_0$ is unfair to H. The second is that $P(\chi > \chi_0 \mid \bar{H}) < 1$, so that the factor from the evidence "$\chi > \chi_0$" is
$$P(\chi > \chi_0 \mid H)/P(\chi > \chi_0 \mid \bar{H}) > P(\chi > \chi_0 \mid H) = P(\chi_0^2).$$

hardly be exactly equal. In fact, if n is very large the problem of estimation of the chances would be more to the point than the problem of significance. A similar remark applies to many other problems and to other tests of significance. (Cf. the remarks at the beginning of 7.8 concerning the meaning of "loaded".)

The difficulties of this example are fairly typical in statistics. Serious mistakes can be avoided only by having a familiarity with the principles of probability.

A question that has been much discussed in recent years is whether it is ever possible to test a hypothesis H by considering its likelihood, but without considering the likelihood of $\bar{H}$. The chi-squared test in its ordinary form does just this. It does not tell us anything immediate about the final odds of H. What it does tell us is that if a statistician always uses the chi-squared test and rejects H when $\chi \geqslant \chi_1$, then he will reject true hypotheses in roughly a proportion $P(\chi_1^2)$ of cases, in the long run. In other words he will commit errors of the first kind in this proportion of cases when H is true—always assuming that the hypotheses that are tested are independent of one another.

If the statistician takes more evidence into account he may be expected to get better results than if he relies on the chi-squared test. But this test often saves time. The saving of time is worth while in any application that is either urgent or not exceptionally important.

7.8B Additional note on the chi-squared test

Let a sample of n objects be classified in terms of ρ mutually exclusive properties; and let the objects fall into ρ cells, the numbers in the cells being $m_1, m_2, \ldots, m_\rho$. Let the (unknown) chances of falling into the cells be $p_1, p_2, \ldots, p_\rho$. On a hypothesis H let $p_1 = \pi_1, p_2 = \pi_2, \ldots, p_\rho = \pi_\rho$, and on the hypothesis $\bar{H}$ suppose that the distribution of the chances is uniform in the space
$$\Sigma p_r = 1, \quad p_r \geqslant 0 \quad (r = 1, 2, \ldots, \rho),$$
with the point $p_1 = \pi_1, p_2 = \pi_2, \ldots, p_\rho = \pi_\rho$ removed. (The notation "$\bar{H}$" is justifiable as in the third paragraph of 6.3 or the eighth paragraph of 7.4.) The square of the "volume" of this space is ρ times the square of the volume of the space
$$\Sigma p_r \leqslant 1, \quad p_1 = 0, \quad p_2 \geqslant 0, \quad p_3 \geqslant 0, \ldots, p_\rho \geqslant 0,$$
by a generalisation of Pythagoras's theorem. (This can be expressed in purely analytical terms, but it is intuitively simpler to use geometrical language.) Hence the volume is $\sqrt{\rho}/(\rho-1)!$, so that the function analogous to ψ in 7.8 is $(\rho-1)!/\sqrt{\rho}$. The factor in favour of $\bar{H}$ is, as in 7.8,
$$\int_{\Sigma p_r = 1} \psi \cdot \prod \left(\frac{p_r}{\pi_r}\right)^{m_r} d\tau,$$

7.8B   PROBABILITY AND WEIGHING OF EVIDENCE

where $d\tau$ is an element of the $(\rho-1)$-dimensional volume. This can be written

$$\frac{(\rho-1)!}{\sqrt{\rho}} \iint \cdots \int_{p_1+p_2+\ldots+p_{\rho-1}<1} \prod_{r=1}^{\rho-1} \left(\frac{p_r}{\pi_r}\right)^{m_r} \left(\frac{1-p_1-p_2-\ldots-p_{\rho-1}}{\pi_\rho}\right)^{m_\rho}$$
$$\times \sqrt{\rho}\, dp_1 dp_2 \ldots dp_{\rho-1} = \frac{(\rho-1)!\, m_1!\, m_2!\ldots m_\rho!}{(n+\rho-1)!\, \pi_1^{m_1} \pi_2^{m_2} \ldots \pi_\rho^{m_\rho}},$$

as we may see by using Dirichlet's integral. This expression for the factor in favour of $\bar{H}$ is exact and can be calculated by means of tables of factorials. By using Stirling's formula we can see that the approximate plausibility gained by $\bar{H}$ is, in natural bels,

$$\tfrac{1}{2}\chi^2 + \log\left\{(2\pi)^{\frac{1}{2}\rho} n^{\frac{1}{2}-\frac{1}{2}\rho} (\pi_1 \pi_2 \ldots \pi_\rho)^{\frac{1}{2}} (\rho-1)^{\rho-\frac{1}{2}} \left(1+\frac{\rho-1}{n}\right)^{-(n+\rho-1)}\right\},$$

where

$$\chi^2 = \sum_{r=1}^{\rho} \frac{(m_r - \pi_r n)^2}{\pi_r n}.$$

The gain in plausibility may be difficult to calculate for other assumptions about the distribution of the chances. In order to get round this difficulty you could frame the body of beliefs in terms of the distribution of χ^2 itself, given $\bar{H}$. The distribution of χ^2 given H is known,† and thus the factor in favour of $\bar{H}$ could be obtained. It should be noticed that the distribution of χ^2 given H is effectively independent of n, whereas the distribution given $\bar{H}$ does depend on n. In fact the expected value of χ^2 given $\bar{H}$ would be an increasing function of n, and the probability density at a fixed value of χ^2 would be a decreasing function of n for large enough values of n. Thus the weight of evidence in favour of $\bar{H}$ for a given value of χ^2 is (for large n) a decreasing function of n, just as it was before. In this respect the method of inverse probability differs from the so-much-or-more method.

For some problems it may not be easy to make a tolerably precise judgment concerning the distribution of χ^2 given $\bar{H}$ or concerning the distribution of the chances given $\bar{H}$. For example, suppose that a die has been bought at a reputable firm and that the spots have been painted on instead of being scooped out, in order that the symmetry should be disturbed very little. It is decided to test the hypothesis H that the die has been made with extreme care, i.e. that the chances are all " exactly " $\tfrac{1}{6}$. The given information may cause you to select

† The probability density of $\zeta = \chi^2$, given H, is very nearly
$$2^{-\frac{1}{2}\nu} e^{-\frac{1}{2}\zeta} \zeta^{\frac{1}{2}\nu-1}/\Gamma(\tfrac{1}{2}\nu),$$
where $\nu = \rho - 1$. The expected value of χ^2 is ν. (See any modern treatise on mathematical statistics.)

for $\bar{H}$ a hypothesis different both from the previous one of the present section (with $\rho = 6$) and from the one in section 7.8. Suppose that $\bar{H}$ is selected in such a way that when it is given the chances are uniformly distributed in a space S' defined by

$$\Sigma p_i = 1, \quad \Sigma(p_i - \tfrac{1}{6})^2 < k^2,$$

for some k between 0·01 and 0·02. (A modification may be desired if the sample frequencies lie too far outside S'.) The arbitrary nature of $\bar{H}$ is justified by the vagueness of the given information. Such vagueness is quite common in the questions which arise in statistics, and this is one of the reasons for the difficulties of the subject.

The "volume" of S' is, as a matter of fact, $8\pi^2 k^5/15$, which is $64\pi^2 k^5/\sqrt{6}$ times the volume of S (with $\rho = 6$). The effect is to increase the plausibility gained by $\bar{H}$, above the value obtained previously, by between 60 db and 75 db.

If n is equal to six million the plausibility gained by $\bar{H}$ is between

$$(2 \cdot 17\chi^2 - 75) \text{ db} \quad \text{and} \quad (2 \cdot 17\chi^2 - 60) \text{ db}.$$

If the initial odds of $\bar{H}$ are between 0·1 and 10, the final plausibility is between

$$(2 \cdot 17\chi^2 - 85) \text{ db} \quad \text{and} \quad (2 \cdot 17\chi^2 - 50) \text{ db}.$$

In order to be able to deduce from this that the final odds of $\bar{H}$ are at least 100 to 1 on we need

$$2 \cdot 17\chi^2 - 85 > 20, \quad \text{i.e. } \chi^2 > 48.$$

To deduce that the final odds of H are at least 100 to 1 on we need

$$2 \cdot 17\chi^2 - 50 < -20, \quad \text{i.e. } \chi^2 < 14.$$

These results may be contrasted with the so-much-or-more method. For instance, given H, the probability that $\chi^2 > 15$ is only 0·001, and such values of χ^2 would normally be regarded as sufficient to reject H. But the discrepancy between the methods is not as large as it seems, since values of χ^2 between 15 and 48 would not be likely to occur, given either H or $\bar{H}$.

The above calculations could easily be modified in order to decide between hypotheses H and $\bar{H}$ where H and $\bar{H}$ are similar to the previous $\bar{H}$ but with associated spaces defined by the inequalities

$$\Sigma(p_i - \tfrac{1}{6})^2 < k_1 \quad \text{and} \quad k_2 < \Sigma(p_i - \tfrac{1}{6})^2 < k_3 \quad (k_1 \leqslant k_2 < k_3).$$

This formulation of the problem corresponds closely to the practical meaning of the question "has the die been made with extreme care?" The vagueness of the question is matched by the fact that k_1, k_2 and k_3 require to be given definite values in order to get a definite answer.

7.9 Contingency tables

The necessity for relying on your own judgment is particularly clear in connexion with the problem of independence in a contingency table. E. S.

7.9   PROBABILITY AND WEIGHING OF EVIDENCE

Pearson and G. A. Barnard have discussed the 2×2 contingency table from this point of view, though not in terms of inverse probability. (See **7.1**.)

We begin with a description of the problem.

Suppose that a population of individuals can be classified with respect to two different properties A and B, e.g. colour of eyes and colour of hair. Let the sub-classes corresponding to these classifications be $A_1, A_2, \ldots, A_r$ and $B_1, B_2, \ldots, B_s$.

Suppose that a sample of the population is taken and it is found that there are n_{ij} individuals in both the classes A_i and B_j.† Let $\sum_j n_{ij} = l_i$, $\sum_i n_{ij} = m_j$, $\sum_{i,j} n_{ij} = n$. These numbers, when arranged in a rectangular array, form a contingency table. (See diagram.)

n_{11} n_{12} . . . n_{1s}	l_1
n_{21} n_{22} . . . n_{2s}	l_2
. .	.
. .	.
. .	.
n_{r1} n_{r2} . . . n_{rs}	l_r
m_1 m_2 . . . m_s	n

A question that is often asked is whether the properties A and B are independent, i.e. whether the chance p_{ij} of belonging to both the classes A_i and B_j is expressible in the form $p_i q_j$, where $\Sigma p_i = 1, \Sigma q_j = 1$. ($i = 1, 2, \ldots, r$; $j = 1, 2, \ldots, s$.)

Sometimes the interesting question is whether the properties A and B are in some sense ‡ *approximately* independent, but here we deal only with the question of strict independence. For small samples we may expect the answer to both questions to be about the same. For large samples it is usually more reasonable to consider the " degree of dependence ", so to speak—a problem of estimation rather than significance.

There is no unique solution to the problem of dependence: the solution must depend on the assumed body of beliefs. Three special bodies of belief will be considered. For these it happens to be possible to obtain a simple exact formula for the factor in favour of dependence. In practice every

† i.e. in the class $A_i.B_j$.
‡ It is not customary to define this sense, so that the question asked is a vague one. (Cf. the remarks concerning vagueness in **7.8B**.) .

example should be treated on its merits, unless the statistician is short of time, and then a rule of thumb like the chi-squared test may legitimately be applied. The way in which this can be done will also be described.

Consider the following six statistical hypotheses, in each of which it is understood that there is a uniform density for the chances within the Euclidean spaces † defined. In all six cases H is supposed to represent the hypothesis that the properties A and B are independent. The "given" propositions, which are not stated, would include descriptions of how the samples were selected. These would probably be different for the three bodies of belief $\mathcal{B}_1$, $\mathcal{B}_2$ and $\mathcal{B}_3$. (It is immaterial whether $\mathcal{B}_1$, $\mathcal{B}_2$ and $\mathcal{B}_3$ are compatible with one another, but if the six hypotheses were all given different symbols then they could be regarded as statistical hypotheses all belonging to the same body of beliefs.)

$\mathcal{B}_1$, $\bar{H}: \Sigma p_{ij} = 1$.
$\quad H: p_{ij} = p_i q_j$, $\Sigma p_i = 1$, $\Sigma q_j = 1$.

$\mathcal{B}_2$, $\bar{H}: \sum_j p_{ij} = l_i/n \ (i = 1, 2, \ldots, r)$ $\Big\}$ where the numbers l_i are known.
$\quad H: p_{ij} = p_i q_j$, $\Sigma q_j = 1$, $p_i = l_i/n$

$\mathcal{B}_3$, $\bar{H}: \sum_i p_{ij} = m_j/n \ (j = 1, 2, \ldots, s)$ $\Big\}$ where the numbers m_j are known.
$\quad H: p_{ij} = p_i q_j$, $\Sigma p_i = 1$, $q_j = m_j/n$

It is not claimed that any of these bodies of belief is "right". They correspond roughly to the cases in which the sampling is done in such a way that

(i) a knowledge either of the column totals only or of the row totals only is felt to affect the probability of independence;

(ii) a knowledge of the row totals is felt not to affect the probability of independence;

(iii) a knowledge of the column totals is felt not to affect the probability of independence.

Now with the help of the mathematical formula

$$\text{Average} \prod_{\nu=1}^{N} x_\nu^{n_\nu} = \frac{(N-1)!\Pi n_\nu!}{(N-1+\Sigma n_\nu)!},$$
$$x_\nu \geq 0, \ \Sigma x_\nu = 1$$

which is connected with Dirichlet's integral, we can prove that the factor in favour of $\bar{H}$, corresponding to $\mathcal{B}_1$ is f, say, where

$$f = \frac{(rs-1)!(n+r-1)!(n+s-1)!\Pi n_{ij}!}{(n+rs-1)!(r-1)!(s-1)!\Pi(l_i!m_j!)},$$

† It will always be taken for granted that the numbers p_{ij} are positive.

and corresponding to $\mathfrak{B}_2$ it is
$$f' = \frac{(n+s-1)!\,\{(s-1)!\}^{r-1}\Pi n_{ij}!}{\Pi(l_i+s-1)!\,\Pi m_j!}.$$
The factor corresponding to $\mathfrak{B}_3$ is similarly
$$f'' = \frac{(n+r-1)!\,\{(r-1)!\}^{s-1}\Pi n_{ij}!}{\Pi(m_j+r-1)!\,\Pi l_i!}.$$
Notice the check that f, f' and f'' all reduce to 1 if $n = 0$ or $n = 1$.†

The reader is recommended to compare these formulae, for the case $r = s = 2$, with those given in standard textbooks on statistics. The factors can all be calculated exactly, or approximated as in **7.8B** by expressions involving χ^2, where
$$\chi^2 = \sum \frac{(n_{ij} - l_i m_j/n)^2}{l_i m_j/n}.$$
Modifications could be made in the various bodies of belief, analogous to those in **7.8B**.

The standard method of applying the chi-squared test to a contingency table is to argue as follows. " If all the numbers l_i and m_j were known this would provide very little evidence about independence. But if these numbers are known and the frequencies l_i/n and m_j/n are identified with p_i and q_j, then (on the hypothesis of independence) the distribution of χ^2 is the usual χ^2 distribution with $(r-1)(s-1)$ 'degrees of freedom'. The appropriate column in the χ^2 tables can then be used in order to find the probability of obtaining a χ^2 exceeding the observed value."

As in **7.8A** the body of beliefs might be formulated in terms of the distribution of χ^2 given $\bar{H}$. The judgments made would depend a great deal on your familiarity with such problems.

Our solutions are not offered in an authoritative spirit, but merely as contributions to a difficult problem. The theoretical difficulties become less acute for large samples. For if r and s are fixed, if n tends to infinity, and if the ratios of $l_i : m_j : n$ are bounded for all i and j, then it is easily seen that the ratios $f : f' : f''$ are also bounded. Hence the alternative judgments will generally all lead to the same decision as to dependence or independence when the sample is very large. But on the chi-squared test the table will nearly always show a significant degree of dependence if n is sufficiently large, for *absolute* independence is rare in real life. This is a theoretical objection to the chi-squared test: you often ask whether the qualities A and B are independent when you really know all the time that they can hardly be absolutely independent. The trouble with the chi-squared test is that it takes the question too literally. (Much the same criticism of the chi-squared test has already been made in **7.8A**.)

† If row and column totals are all irrelevant the factor may reasonably be taken as $f(f'/f)(f''/f) = f'f''/f$.

STATISTICS AND PROBABILITY

One method of using f, f', f'' is to calculate them and then to use the results as a basis for further judgment. The calculation of f, f' and f'' is objective, so that the method is similar to the use of the chi-squared test. The results at least serve as a check on the reliability of the chi-squared test.

The formulae for f, f' and f'' bear a formal resemblance to the likelihoo ratio † λ for the hypothesis of independence. λ is easily seen to be ‡

$$\lambda = \frac{\prod l_i^{l_i} \prod m_j^{m_j}}{n^n \prod_{i,j} n_{i,j}^{n_{ij}}}.$$

This formal resemblance should not be taken to imply that λ can be given an interpretation similar to that of a factor. In fact λ cannot exceed unity. λ is used by considering its distribution on the assumption of independence, whereas the factors can be interpreted directly.

7.10 Estimation problems

We shall now consider the problem of the estimation of the values of a set of unknown numbers. For simplicity, however, it will be supposed that there is only one number c, though everything that will be said can be extended to any finite set. Some examples of estimation have already been discussed. The problem is to associate with c either a "best" value or a whole interval of values. Here we shall deal only with the latter problem.

An important case is when c is the only parameter in a composite statistical hypothesis H, so that H is the disjunction of simple statistical hypotheses H_c for some class of real values of c. Let E be "an experiment", i.e. a collection of statistics. (See the first footnote in **7.4**.)

It is generally agreed that if the initial distribution of c is known then the final distribution can be obtained, and the probability that c will lie in a given interval can be deduced at once. But usually the initial distribution of c is not known precisely, being only partly defined by means of inequalities. The question arises then whether anything "precise" can be said about c, i.e. anything that does not depend on the initial distribution. In fact this can be done in the following ingenious way.

Let $\underline{c}(E)$ and $\bar{c}(E)$ be numerical functions of E. Suppose that for all c and some fixed α,

$$P\{\underline{c}(E) \leqslant c \leqslant \bar{c}(E) \mid H_c\} = \alpha.$$

Then the interval $[\underline{c}(E), \bar{c}(E)]$ is called a *confidence interval* for c with *confidence coefficient* α.

---

† This is defined, for example, by S. S. Wilks, 1944, 150. The likelihood ratio should not be confused with the ratio of the likelihoods used in the definition of a factor. Wilks's definition, slightly generalised, is given in a footnote in our Section **6.1**.

‡ Wilks, in error, gives the value of λ^{-1}. (*L.c.*, 220.)

7.10 PROBABILITY AND WEIGHING OF EVIDENCE

It should be carefully noticed that the "given" evidence in the above probability is H_c, although in practice it is E which is known and not H_c.

Now suppose that the functions $\underline{c}(E)$ and $\bar{c}(E)$ are selected so that $[\underline{c}(E), \bar{c}(E)]$ is a confidence interval with coefficient α, where α is near 1. Let us imagine that the following instructions are issued to all statisticians.

"Carry out your experiment, calculate the confidence interval, and *state* that c belongs to this interval. If you are asked whether you 'believe' that c belongs to the confidence interval you must refuse to answer. In the long run your assertions, if independent of each other, will be right in approximately a proportion α of cases." (Cf. Neyman (1941), 132–3.)

The advantages and disadvantages of the procedure are similar to those of the chi-squared test and hardly require additional comment. We remark merely that if the procedure were consistently adopted it would occasionally lead to ridiculous behaviour, because of its neglect of initial probabilities and utilities.

A technique that bears some resemblance to that of confidence intervals is that of "tolerance limits". (See Wilks (1946).) Suppose that X is a continuous random variable with an unknown density function $f(x)$. A sample of n independent readings is selected and these are arranged in numerical order $x_1 \leqslant x_2 \leqslant x_3 \leqslant \ldots \leqslant x_n$. Let $L_1(x_1, x_2, \ldots, x_n)$, $L_2(x_1, x_2, \ldots, x_n)$ be two functions of the sample values. These functions are called "$100\beta\%$ distribution-free tolerance limits at probability level α" if, whatever function f may be,†
$$P\left(\int_{L_1}^{L_2} f(x)\,dx \geqslant \beta\right) = \alpha,$$
assuming that the probability density of X is $f(X)$. In particular, Wilks shows that $L_1 = x_1$, $L_2 = x_n$ are such tolerance limits if
$$n\beta^{n-1} - (n-1)\beta^n = 1 - \alpha.$$
For example, if $n = 473$, it is 19 to 1 on that the interval $[x_1, x_n]$ will include at least 99 per cent of the population. But this is true only before the sample is selected. Afterwards it is likely to be more informative to take all the readings into account and to use a curve-fitting technique, *even if the curve-fitting is done by eye*.

Thus the technique of tolerance limits is liable to throw away evidence for the sake of objectivity. In this it again resembles the chi-squared test, and like the chi-squared test its convenience depends partly on whether suitable tables are available.

The importance of these objective techniques should not be underestimated. By ignoring subjective judgments they are incapable of giving information about the final probabilities of the hypotheses, but they do give results that are indisputable and they often give them without much calculation.

† Observe that the *existence* of f is assumed.

STATISTICS AND PROBABILITY 7.10

The general conclusion is that in statistics it is useful to know a number of different techniques, the basic one being the technique of probability.

Exercise. An "unbiased estimate" of a parameter c is a statistic whose expected value, given c, is c. In a sequence of n independent trials with chances p there are r successes. Show that an unbiased estimate of p^k is $r^{(k)}/n^{(k)}$ where $s^{(k)} = s(s-1)(s-2) \ldots (s-k+1)$. This actually vanishes if $r < k \leqslant n$. Assuming that p has a uniform initial distribution show that the expected value of p^k is $(r+k)^{(k)}/(n+k+1)^{(k)}$.

APPENDICES

I. The error function

Several books on probability include tables of the "error function". Here we content ourselves with the following approximate formula for mental calculations:—

$$- 10 \log_{10} \frac{1}{\sqrt{2\pi}} \int_x^\infty e^{-\frac{1}{2}t^2} dt = 2\tfrac{1}{6}x^2 + 4 + 10 \log_{10} x,$$

with an error less than 1 if $2 \leqslant x \leqslant 14$.

II. Dirichlet's multiple integral †

$$\int\int \ldots \int x_1^{m_1} \ldots x_n^{m_n} f(\Sigma x) dx_1 \ldots dx_n$$
$$= \frac{m_1! \ldots m_n!}{(\Sigma m + n - 1)!} \int_0^a f(x) x^{\Sigma m + n - 1} dx,$$

where the region of integration in the multiple integral is defined by $\Sigma x \leqslant a$, $x_1 \geqslant 0, \ldots, x_n \geqslant 0$. The formula is not restricted to integral values of m_1, $m_2, \ldots$, but these numbers must be algebraically large enough for the integrals to exist.

It can be deduced that the volume of an n-dimensional unit sphere is $\pi^{\frac{1}{2}n}/(\frac{1}{2}n)!$, a result which was used in 7.8B.

III. On the conventionality of the addition and product laws ‡

We shall show (but not quite rigorously, nor in detail) that the addition law for mutually exclusive "events" and the product law for independent events are largely conventional. At first this appears to exhibit an essential distinction between the non-frequency and frequency theories. But it should be realised that in the frequency theory it *is* likewise only a convention to define probability as the limit of a proportion of successes rather than as some monotonic function of this limit.

Suppose that "probability$_A$" (denoted for short by P_A) has the properties—

(i) $P_A(E.F)$ is a function of $x = P_A(E)$ and $y = P_A(F)$ where E and F are arbitrary independent events. (We are taking the "given" proposition for granted.)

(ii) $P_A(E \vee F)$ is a function of $P_A(E)$ and $P_A(F)$ where now E and F denote mutually exclusive events.

Since $E.F \equiv F.E$ and $E.(F.G) \equiv (E.F).G$, with similar results for disjunctions, it follows that the two functions mentioned satisfy the commutative

---

† See, for example, Whittaker and Watson, *Modern Analysis* (4th edn., 1927), 258, or Jeffreys and Jeffreys, *Methods of mathematical physics* (1946), 440.

‡ The following remarks arose out of a discussion on a paper by G. A. Barnard (*Jour. Roy. Stat. Soc.*, Ser. B, 1949 or 1950) and many of the ideas are his. See also "conventions" in the Index, for references to Jeffreys and Schrödinger.

APPENDICES

and associative laws. It then follows from a theorem † due to Abel (and published in his collected works) that the two functions are of the forms

$$\varphi^{-1}\{\varphi(x) + \varphi(y)\}, \quad \psi^{-1}\{\psi(x) + \psi(y)\}.$$

Now define $P_B(E)$ as $\exp \varphi(P_A E)$. Then P_B satisfies the product law and a modified addition law of the form

$$\tau(t) = \tau(x) + \tau(y),$$

where $x = P_B(E)$, $y = P_B(F)$ and $t = t(x, y) = P_B(E \vee F)$. Now

$$(E.F) \vee (E.G) \equiv E.(F \vee G),$$

so the function $t(x, y)$ satisfies the condition of homogeneity

$$t(\lambda x, \lambda y) = \lambda t(x, y).$$

It can be deduced from these conditions that the function t is of the form $(x^K + y^K)^{1/K}$, for some constant K. (This is not a trivial result. It is necessary to assume at least that the function is measurable.) Now, at last, let probability be defined by $P(E) = (P_B E)^K$. Then probability satisfies the product law and the ordinary addition law. Thus it is sufficient to assume quite weak properties for probability$_A$ in order to establish the existence of a probability which satisfies the addition and product laws. Moreover, probability is an increasing function of probability$_A$ since exponentials and Kth powers are increasing functions. Therefore the partial ordering for probability is the same as for probability$_A$. This shows in what sense the addition and product laws are conventional.

† This theorem gives necessary and sufficient conditions for a function of two variables to be calculable on a suitably calibrated slide-rule. The theorem has been rediscovered several times. See, for example, J. Aczél, *Bull. Soc. math. Fr.*, **76** (1948), 59–64.

REFERENCES

BARNARD, G. A., 1946. Sequential tests in industrial statistics. *Journ. Roy. Stat. Soc., Supplement*, **8,** 1–21. Discussion, 22–6.
——, 1947. Significance tests for 2 × 2 tables. *Biometrika*, **34**, 123–38.
BARTLETT, M. S., 1933. Probability and chance in the theory of statistics. *Proc. Roy. Soc.*, A, **141**, 518–34.
——, 1936. Statistical probability. *Journ. Amer. Stat. Ass.*, **31**, 553–5.
——, 1940. The present position of mathematical statistics. *Journ. Roy. Stat. Soc.*, **103**, 1–19.
——, 1946. The large sample theory of sequential tests. *Proc. Camb. Phil. Soc.*, **42**, 239–44.
BRUNT, D., 1931. *The combination of observations.* Cambridge. 2nd edn.
CRAMÉR, H., 1937. *Random variables and probability distributions.* Cambridge.
——, 1946. *Mathematical methods of statistics.* Princeton.
——, 1947. Problems in probability theory. *Annals of Math. Stat.*, **18**, 165–93.
ELDERTON, W. P., 1938. *Frequency curves and correlation.* Cambridge.
FELLER, W., 1945. The fundamental limit theorems in probability. *Bull. Amer. Math. Soc.*, **51**, 800–32.
FISHER, A., 1922. *The mathematical theory of probabilities and its application to frequency curves and statistical method.* 2nd edn., New York.
FISHER, R. A., 1938. *Statistical methods for research workers.* Edinburgh and London.
FRÉCHET, M., 1937. *Généralités sur les probabilités : variables aléatoires.* Paris.
HALDANE, J. B. S., 1931. A note on inverse probability. *Proc. Camb. Phil. Soc.*, **28**, 55–61.
HILBERT, D., and ACKERMANN, W., 1946. *Grundzüge der theoretischen Logik.* 1st edn., Berlin, 1928 ; 2nd edn., 1937 ; reprint New York, 1946.
JEFFREYS, H., 1936. Further significance tests. *Proc. Camb. Phil. Soc.*, **32**, 416–45.
——, 1937. *Scientific inference.* Cambridge.
——, 1939. *Theory of probability.* Oxford.
——, 1942. Probability and quantum theory. *Phil. Mag.*, **33**, 815–31.
——, 1946. An invariant form for the prior probability in estimation problems. *Proc. Roy. Soc.*, A, **186**, 453–61.
JESSEN, B., and WINTNER, A., 1935. Distribution functions and the Riemann zeta function. *Trans. Amer. Math. Soc.*, **38**, 48–88.
KEMBLE, E. C., 1942. Is the frequency theory of probability adequate for all scientific purposes ? *Amer. Journ. Physics*, **10**, 6–16.
KENDALL, M. G., 1945. *The advanced theory of statistics*, Volume 1. 4th edn., 1948, London. Volume 2 appeared in 1946 (2nd edn., 1947.)
KEYNES, J. M., 1921. *A treatise on probability.* London.
KOLMOGOROFF, A., 1933. *Grundbegriffe der Wahrscheinlichkeitsrechnung.* Berlin.
KOOPMAN, B. O., 1940. The basis of probability. *Bull. Amer. Math. Soc.*, **46**, 763–74.

REFERENCES

KOOPMAN, B. O., 1940. The axioms and algebra of intuitive probability. *Annals of Math.*, **41**, 269–92.
MISES, R. von, 1936. *Probability, statistics and truth.* London. Original German editions, 1928 and 1936. Vienna and Berlin.
——, 1942. On the correct use of Bayes's formula. *Ann. Math. Stat.*, **13**, 156–65.
——, 1945. *Wahrscheinlichkeitsrechnung.* New York. Originally Leipzig-Vienna, 1931.
NEYMAN, J., 1941. Fiducial argument and the theory of confidence intervals. *Biometrika*, **32**, 128–150.
NEYMAN, J., and PEARSON, E. S., 1933. On the testing of statistical hypotheses in relation to probability *a priori*. *Proc. Camb. Phil. Soc.*, **29**, 492–510.
—— ——, 1933. On the problem of the most efficient tests of statistical hypotheses. *Phil. Trans.*, A, **231**, 289–337.
PEARSON, E. S., 1938. The probability integral transformation for testing goodness of fit and combining independent tests of significance. *Biometrika*, **30**, 134–48.
——, 1942. Notes on testing statistical hypotheses. *Biometrika*, **32**, 311–16.
——, 1947. The choice of statistical tests illustrated on the interpretation of data classed in a 2 × 2 table. *Biometrika*, **34**, 139–67.
POINCARÉ, H., 1912. *Calcul des probabilités.* Paris.
RAMSEY, F. P., 1931. *The foundations of mathematics.* London.
REICHENBACH, H., 1932. Axiomatik der Wahrscheinlichkeitsrechnung. *Math. Zeitschrift*, **34**, 568–619.
SCHRÖDINGER, E., 1947. The foundation of probability. *Proc. Roy. Irish Acad.*, **51A**, 51–66 and 141–6.
TODHUNTER, I., 1865. *A history of the mathematical theory of probability.* Cambridge and London.
USPENSKY, J. V., 1937. *Introduction to mathematical probability.* New York.
VENN, J., 1888. *The logic of chance.* 3rd edn., London.
WALD, A., 1945. Sequential method of sampling for deciding between two courses of action. *Journ. Amer. Stat. Assoc.*, **40**, 227–306.
——, 1945. Sequential tests of statistical hypotheses. *Ann. Math. Stat.*, **16**, 117–86.
—— 1947. *Sequential analysis.* New York.
WILKS, S. S., 1944. *Mathematical statistics.* Princeton.

INDEX

A few definitions and remarks are included for the sake of clarity.
The references on pages 107-8 have not been indexed.

A

A1 to A6, 19
A4', 49
Abel, N. H., 105
abstract theory, 5, 19–30
acceptance, 65, 84
Ackermann, W., 1n, 27n
acoustics, 63, 64
actuarial work, 53
Aczél, J., 105n
addition law, 13, 16, 19(A2), 104
    generalised, 22–3, 26, 27
    see additivity, complete
addition of random variables, see sum
additivity, complete, 5n, 23, 29, 50n
adultery, 74
almost certain, 18, 21, 26, 27, 39, 46, 52
    see certain
almost certain (or impossible) and empirical propositions, 35, 78
almost certain, and infinite successions of trials, 29
almost impossible, 18, 21
    propositions " given ", 30, 39–40, 46n
    see impossible; almost certain
almost mutually exclusive, 21
alternative hypotheses or theories, 40–6, 64–6, 99
alternatives, 14
and, 1
approximation, 33, 34, 36, 37n, 46, 49, 51, 56, 59, 60, 69, 81, 88, 90, 91, 92, 93, 98, 104
asymptotic properties, 77
attributes, 76, 77, 78, 85
authority, 12, 100
average = arithmetic mean. Not to be confused with " mean "

average, as a maximum likelihood value of a normally distributed variable, 87
axiom, additional, see additivity, complete
axiom, alternative, 49
axiomatic method, 5
axioms, see A1, etc.
    alternative set, 21, 30
    " obvious ", 13, 20, 53
    of logic and mathematics, see H^*
    of utility, 53
    origin of, 13–18
    rules and suggestions, 12, 31, 34, 47

B

$\mathcal{B}$, see body of beliefs
$\mathcal{B}^*$, 47
$B(E \mid H)$, 2
Barnard, G. A., 64n, 77n, 98, 104n
Bartlett, M. S., 10n, 11, 42, 73, 77n
Bayes' postulate, 9n, 55
    see insufficient reason
Bayes' theorem, 24, 40, 62, 63, 65, 67, 68, 71, 77, 94
    see probability, inverse
Bayes' theorem in reverse, 35, 81
    see imaginary results, device of
bel, 63
    natural, 63n
belief, see degrees of belief
beliefs, body of, see body of beliefs
benefit (expected), see utility
Bernoulli, Daniel, 54
Bernoulli, Jacob, 6n, 29n
" best " value of a parameter, 101
betting, see gambling
bias, 41, 45, 89n
    see dice, loaded; unbiased estimate

109

INDEX

billiard balls, 9
binary digit, 75
biology, 83
   see genetics
Birkhoff, G., 14n
birthdays, 38
" bit " of information, 75
blood-groups, 74
body of beliefs—
   alternative, 43, 99
   augmentation, 32
   definition, 3, 32
   empty, 4
   for a contingency table, 99, 100
   generalisation of, 10, 48–9
   taken for granted, 20
   transitive, 14n
Boltzmann's constant, 75
Borel's theorem (perhaps better called the *Borel-Cantelli theorem*), 29, 46, 78
brackets, 26n
Broad, C. D., 21n
Brunt, D., 90
" bumpiness ", 85, 86

C

calculation, *see* numerical work
Cantelli, F. P., 29n
   see Borel
cards, 8, 34, 37, 38, 73
   perfect, perfectly shuffled, 15, 16 34
Carnap, R., 11n, 48
Cauchy-Schwartz inequality, 39
causes, 60
central limit theorem, 57, 88
certain(ty), 19, 21, 24(T7)
   see almost certain
   practical, 6, 39, 49
chance, 41, 78, 82, 84
   and sampling, 78, 79
   distribution of, 79–82, 84
   expectation of, 79
   games of, *see* games of chance
   maximum likelihood value, 80

chance, probability of, *see* probability of a chance
   " true ", 43, 46n
chances, classification of, 43
characteristic function, 54, 59
   discrete, 58
cheating, 44n
chemistry, 76
chess, 49
chi-squared test, 70, 77, 84, 92, 93–7
   analogy with confidence intervals, 102
   analogy with tolerance limits, 102
   and contingency tables, 99, 100
   formula for distribution, 96n
chromosomes, *see* genetics
class interval, 59
classical definitions (of probability), 35
cogent reason, 8, 12, 37, 47
coin-spinning, 36–7, 43, 47, 53, 72, 75
collective, 7
common sense, 67, 77
comparable degrees of belief, 3, 9, 13
comparison between beliefs, 3, 11, 13–14, 32, 33, 37
complication, 36, 76
compounding of distributions, *see* convolution; sum of random variables
computable numbers (for a definition *see* Turing, *Proc. London Math. Soc.*, 1937), 55n
conditioned reflexes, 7
confidence coefficient, 101
confidence intervals, 77, 101–2
conjunction, 1
   · *see* multiplication law
consistency of the abstract theory, 5, 21, 30, 33
constructibility, 4n, 32n
contingency tables, 77, 97–101
continuity, *see* mathematical convenience
contradiction, 3, 20, 21
convenience, *see* mathematical convenience
conventions, 9n, 13, 15, 104
convolution, 52, 56, 57, 88

INDEX

Copeland, A. H., 7n
correlation coefficient, 58
cosmic rays, 82
Coxeter, H. S. M., 38
Cramér, H., 9, 23n, 50n, 51n, 57, 76n
credibility, 2n
crime, *see* law (legal)
cumulants, 59
curve-fitting, 84–9, 102
curves—
    freehand, 89, 102
    J- and U-shaped, 88
    single and double humped, 86, 87, 88
    see smoothness ; " bumpiness "

D

Davenport, H., 38
db, *see* decibel
decibel, 63–4
decimals, 17–18, 57, 93
definitions, 19, 21, 30
    see under probability *and other headings*
degrees of belief, 1–3
    concerning mathematical theorems, 49
    sometimes meaningless, 2, 3n, 30, 32
    see comparison ; intensity ; probability
degrees of dependence, 98
degrees of freedom, 100
degrees of meaning, 1n, 40n
density function, 51, 54, 93
dependence, *see* independence
determinism, 15
dice (imperfect), 38, 59, 80, 96–7
    loaded, 64, 67, 72, 90–4
    perfect, 16, 17
digits, 58, 75
dimensions, 7n
    see space, finite-dimensional
Dirichlet's multiple integral, 74, 96, 99, 104
dishonesty, 45
    see honesty
disjunction, 1
    see addition law
distances, *see* geometry

distribution, 50–61
    binomial, 28, 56, 57
    -free, 102
    frequency, 59–61
    function, 50
    neither continuous nor discrete, 55
    normal, 56, 57, 60, 86–90, 104
    " observed ", 88
    of a chance, 79–82, 84
    of a distribution, 84
    Poisson, 56
    rectangular, 18, 55, 56, 69, 70, 81, 95, 97
    and contingency tables, 99
    and Laplace's law of succession, 80
    and maximum likelihood, 83
    two-dimensional, 51, 58
distributions—
    compounding of, 51, 54, 88
    linear combination of, 88
    see curves
dogs, conditioning of, 7
Dreyfus, A., 67
Drosophila, 82
dualism, *Preface*, 11n, 42n

E

E, 1, 2, 2n, 82
E^*, 19
economics, 83
Elderton, W. P., 88n
electronic reasoning, 48
elementary symmetric function, 38n
entropy, 75
equally probable cases, 7–8, 13–18, 26, 33, 47
equivalence, 19, 29n
error, avoidable or unavoidable, 89
    function, *see* distribution, normal
errors of the first and second kind, 65n, 77, 83, 84, 94, 95
ESP, *see* extra-sensory perception
estimation, 81, 95, 98, 101–3
ethics, 53n

111

INDEX

"evens", 62
event, 33–4
    rare, 88
evidence—
    circumstantial, 67
    ignoring of, 36, 77, 93, 102
    see weight of evidence ; information
    exclusive, 14, 16, 22
    exhaustive, 14, 26
expectation, 52–4, 58
    For expected odds, etc., *see under separate headings*
experiment, 6, 6n, 8, 75
    see E ; trials
experiments—
    conceivably repeated, 7, 47
    design of, 35–6
extra-sensory perception, 35, 37, 44–5, 66, 68–70, 81, 90
eye-colour, 23, 98

F

factor, 62–4
    bounds for, 68
    expected, 72
    infinite, 67
    large, 68
    maximum, 91
    moments of, 74
    partial, 68, 71
    relative, 71, 79, 90
    sometimes of importance apart from the initial probability, 70n
    used as a statistic, 100–1
    see sequential tests
factors, weighted average, 68, 91
"failure", *see* "success"
fallacy of typicalness, 67
Faltung, 52
Feller, W., 29n, 52n
final (probability), 24, 71–2, 83
finite-frequency theory, 9n
Fisher, Arne, 8n
Fisher, R. A., 36, 62, 63n, 76, 82
forecasting, 49

fractional dimensions, 7n
Fréchet, M., 23n, 29n
frequency, 59–61, 77–8
    limiting, 6, 29, 46, 78, 79n, 82
    theory of probability, 6–7, 29, 46–7, 80
        apparent concession to, 12
    see finite-frequency theory
function space, 61, 85
future and past, 1, 2n

G

gambling (and betting), 49 (*bis*), 53–4, 73
    impossibility of a system, 7
games of chance (idealised), 13, 16, 78
    see cards ; coin-spinning ; dice
Gaussian distribution, *see* distribution, normal
genetics, 41, 70, 71n, 74, 82
geometrical language, 95
geometry, 4, 32
    (distance), 11, 34
"given" proposition unknown, 102
guessing, *see* extra-sensory perception

H

H, 1
H^*, 19–20, 24
Haldane, J. B. S., 55, 56, 63n, 70
happiness, 53
    see utility
Hardy, G. H., 72n
Hausdorff, F., 7n
hearsay evidence, 36
height of men, 43, 59, 84–8
heredity, *see* genetics
Hermite functions, 89
Hilbert, D., 1n, 27n
Holmes, S., 67
honesty, 35, 55
    see dishonesty
hypotheses—
    alternative, *see under* alternative
    considered in pairs, 66, 83–4
    three, 66

INDEX

hypothesis, 24, 40-6
   acceptance and rejection, 84
   composite, 68-70
   plausible, 83
   stated after making observations, 91, 94
   statistical, 66, 73
   statistical, composite, 82, 101
   statistical, simple, 66, 82, 99, 101
hypothetical population, 60n, 78
   see population ; super-population

I

ideal, unattainable, 6
idealised games of chance, *see* games of chance
idealised problems, 5, 15n, 17-18, 35, 80
   see additivity, complete ; probability, infinite ; proposition, idealised
ignorance, 15
ignoring of information, *see* evidence, ignoring of
imaginary results, device of, 35, 70
   see Bayes' theorem in reverse
imaginary universe or world, 1, 42
imagination, 41n
implication, 19
   We usually interpret this as " logical implication ". But all the theorems can be extended to the case of " material implication " by regarding as one of the " given " propositions the proposition H^{**} which asserts all true laws of nature. It will be found convenient sometimes to take H^{**} for granted and omit it from the notation
importance *versus* urgency, 95
impossible, 14, 19, 24
   see almost impossible ; proposition, self-contradictory
improper theories, 41-3, 69
inaccurate language, 33-4
incompatible, *see* mutually exclusive
inconsistent, *see* unreasonable ; consistency
independence, 17, 21-3, 67, 78, 95
   in a contingency table, 98-100

independent random variables, 51
indeterminism, 15
indifference, principle of, 37
   see insufficient reason
individuals, 76
induction, scientific, 11, 41
inequalities, 27, 38-9, 72
   see comparison
inertial constants, analogues of, 58
infinite—
   " approximately ", 7
   expectation, 53-4
   factor, 67
   number of hypotheses, 44, 46n, 69
   number of parameters, 61
   number of propositions, 22
   population, hypothetical, *see* hypothetical population
   probability, 21, 55-6
   succession of trials, 6-7, 18, 29
infinity, *see* mathematical convenience
inflexion, points of, 86, 88
information—
   amount of, 63, 74-5
   half-forgotten, 36
   vague, 66, 97, 98n
   see evidence
initial distribution, insensitivity with respect to, 80
initial probability, etc., 24, 35, 45, 46, 60, 62, 71-2, 83, 84, 101, 102
instructions to statisticians, 102
insufficient reason, 8, 37, 55
   see cogent reason
insurance, 53
intensity of belief, 1-3, 32
   see comparison
intuition, 49, 78
intuitionism, 49
irregular collective, 7

J

Jeffreys, H., 2, 4, 8-9, 11-14, 21, 24n, 35, 42n, 47, 55-6, 60, 63, 104n
Jessen, B., 88

113

Johnson, W. E., 10
judgment, 48–9, 65, 77, 80–1, 84, 85, 89, 100
    see probability judgments
jury, *see* law (legal)
justification (*a priori*), 13, 33
    see verification

K

Kemble, E. C., 8
Kendall, M. G., 57, 76n, 88n
Keynes, J. M., 2, 10, 14n
Khintchine, A., 29n, 52n
Kneale, W., 9n
Kollektiv, 7
Kolmogoroff, A., 9, 23n, 29n, 52n
Koopman, B. O., 3n, 10, 11

L

language—
    design of, 4n, 48
    geometrical, 95
    inaccurate, 33–4
    non-mathematical, 34
    probability depending on, 48
Laplace's law of succession, 80
law—
    (frequency distribution), 60
    (legal), 36, 47, 66–7
    of large numbers, 52n
    of nature, 32, 60n
        see scientific theories
    see addition law ; multiplication law
laziness, 77
Lebesgue, H. L., 7n, 9, 23, 51
    see measure
legal applications of hypotheses, 66
Legendre polynomials, 85n
Lévy, Paul, 29n
likelihood, 62, 83
    maximum, 73 (definition), 77, 80, 82–3, 87, 89
    precise, 82
    ratio, 63n, 101
likely = probable. *But see* likelihood
limit, *see* frequency, limiting

Littlewood, J. E., 72n
logic, 1, 2, 5, 19, 27
    formal, a contrast with probability, 14
    inadequacy of formal, 3
logical notation, 1
logically true, and false, 19
" long run ", 10
" lot ", 64

M

m, 2, 3, 4
mathematical convenience, 18, 36, 51, 60, 79, 88, 94, 102
    see additivity, complete
mathematical theorems, beliefs concerning, 49
mathematics, pure, 19, 49, 76
maximum expected utility, 53
Maxwell demon, 75
mean (or mean value), *see* expectation
mean deviation, 55
mean value of a chance, 79
meaning, 1, 3n, 4n, 5, 40n
    degrees of, 1, 40n
    see under degrees of belief
measure, 7n, 9, 18n, 21, 23
    see function-space
measurement, 50, 89–90
median, 55
medicine, 83
Mendel, G. J., 41
meteorology, 49
miracles, 39
Mises, R. von, 6–7, 10, 24n, 29
mistake, 89, 95
models, 38
moments, 54, 56, 58, 59, 88
money, 53–4
most probable value (a value of a parameter for which the point or density function is a maximum), 80, 83
motive, 67
multiplication law, 13, 16–17, 19(A3), 22, 23, 24, 27(line 3), 104

INDEX

murder, 67
music, 64
mutation, 82
mutually exclusive, 14, 16, 22

N

Nagel, E., 11n
negation, 1, 25
neper, 63
Neyman, J., 77, 83, 94, 102
non-numerical theory, *see* probability, numerical
not, *see* negation
notation—
    ambiguous, 32
    logical, 1
    " misleading ", 17, 21n, 50, 66, 84
numerical work, 55

O

O, o (should not be confused with the same symbols used in pure mathematics for orders of magnitude), 62
objective, constructibly, 4n, 32n
objective (and subjective) degrees of belief, comparisons, probabilities and theories of probability, 2, 4, 6–11, 42, 47–8
    see precision
objectivity—
    and the neglect of evidence, 102
    degrees of, 4
    superficial appearance of, 6
observations, combination of, 89–90
Occam's razor, 60
octave, 63n, 64, 75
odds, 62, 73, 83
    expected, 73
    gambling, 49
opinion, differences of, 83
    see public opinion
or, *see* disjunction
oxygen, 39

P

P and P', 32, 36n
$P(E)$, 21
    see notation, " misleading "
$P(E \mid H)$, etc., 2, 4, 19, 31
parabolas, 89
parameters in a law, 60–1, 85, 101
partial ordering, 14n
past and future, 1, 2n
Pearson, E. S., 77, 82n, 83, 85n, 94, 98
Pearson, K., 88–9
perfect coins, packs of cards, etc., *see* games of chance (idealised); cards, perfect
Petersburg problem, 53
philosophy—
    independence of abstract theory from philosophical interpretation of probability, 29
    solipsism, 11
    see unobservables
physics, 36
    see quantum theory
π, 49
plausibility, 63
    gain or loss, *see* weight of evidence
    levels, 65
    relative, 71
players' ruin, 73
Poincaré, H., 27
point function, *see under* probability
point-set theory, 7n, 9, 21, 23
Poisson distribution, 56
politics, 41n
Pólya, G., 72n
polynomials, *see* Legendre polynomials; parabolas
population (finite or infinite), 59–60, 76, 78, 80, 82, 84, 85n
posterior, *see* final
practical difficulties, 36, 76
practice, closeness of our theory to, 12
precision, 34, 42, 47n, 82–4, 90, 101
    see probability intervals
prediction, 6, 39, 49, 60, 76

115

INDEX

primitive notions (beliefs and comparisons between beliefs), 2
Primula sinensis, 70
prior, *see* initial
probability, 3, 14, 19, 31
    The definition of **1.3** is finally completed in **4.1** where " probability " is given a double meaning. In chapter 2 the word is used in a restricted sense and in chapter 3 without any definite meaning. (The word is also used occasionally instead of " a theory of probability ")
    abstract theory of, 5, 19–30
    ambiguous definition, 9–10
    and language, *see* language
    and statistics, 76–103
    circular definition, 6
    close to one, *see* certainty, practical
    continuous or geometrical, 17–18, 40
    definitions of, 6–12
    density, 51, 54, 93
    distribution(s), *see* distribution
    equal, 14
    expected, 73
    experiments, 8
    final, *see* final
    fundamental theorem of, 46, 78, 81
    geometrical, *see* probability, continuous
    given all known information, 36, 41n
    infinite, 21, 55–6
    initial, *see* initial
    intervals, 40, 82
        see precision
    inverse, 62, 70, 82–4
        see Bayes' theorem
    irrational, 18, 34
    judgments, 3, 4, 12, 14, 49, 61, 67, 82, 94
        see judgment
    linguistic, 48
    non-negative, 19(A1), 25
    numerical, 6, 10, 14–15, 20, 34, 36, 37
    objective, *see* objective

probability of a chance, 43
    see distribution of a chance
    of a distribution, 84
    of a logical combination of propositions, 27
    of E given H, *see* $P(E)$, $P(E \mid H)$
    one, *see* almost certain
    physical, *see* quantum theory ; objective (probabilities)
    point function, 51, 54, 91
    posterior, *see* final
    precise, *see* precision
    prior, *see* initial
    relative, 71, 79
    small, 39, 67, 68
        see rare events
    statements, 19, 20, 41, 42n
        see proposition, definition of
    statistical, 42
    tautological, 42, 82
    " technique ", 31, 33, 103
    theories of, *see* theories of probability
    theory of, *see* theory of probability
    true, *see* chance, " true "
    zero, *see* almost impossible
probability$_1$, 48
product rule, *see* multiplication law
proper and improper (theories), 41–3, 69
proportion of possible alternatives, 9
proposition—
    analytic, 2, 19
    definition of, 1, 3, 19, 20, 41, 42n, 72n
    empirical, 2, 30, 34–5, 78, 90n
    idealised, 90n
    incompletely defined, 42, 82
    involving probability, 1, 19, 20, 41, 42n, 72n
    " partial ", 1n
    self-contradictory, 20
propositional functions, 37n
    (A propositional function is a function whose values are propositions)
propositions, logical combination of, 27
psychology, 7, 11
public opinion surveys, 41n

116

INDEX

"pure thought", 26
Pythagoras's theorem, generalisation, 95

Q

quality control, 64–6
quantum theory, 41–2, 76, 78
   see unobservables
question,
   refusal to answer, 102
   taken *too literally*, 100

R

rain, 1, 36
Ramsey, F. P., 10, 53
random—
   at, 38
   numbers, 57, 58
   sample, 38, 78
   variable, 50
rare events, 88
   see probability, small
rational—
   behaviour, 53
   numbers, as probabilities, 17, 34
reasonable, *Preface*, 2, 3, 9, 33
   see rational behaviour
reasoning—
   definition, 3
   electronic, 48
recognition, 68
Reichenbach, H., 10n
rejection, *see* hypothesis, acceptance and rejection; "lot"; observations, combination of
relevance, 36
resultant, 52
rigour, 76
roulette, 16
rules, 5, 31–2
   see axioms, rules and suggestions
Russell, Bertrand, 2n, 9n, 37n

S

"same essential conditions", 46
sample, 60

sample, frequency, 59–60, 77–8
   mean, etc., 60
   size expected, 65, 73
   small and large, 77, 83, 95, 98
sampling—
   and chance distributions, 84–8
   of a single attribute, 77–81
   with and without replacement, 38, 78–9, 80, 85
scale readings, 50, 89–90
Schrödinger, E., 13, 104n
Schwartz, H. A., 39
scientific mind, 77
scientific theories, *Preface*, 1n, 4, 10, 31, 40–6
   see law of nature
self-consistency, *see* consistency
semitone, 64
sequential tests, 64–6, 73
Shannon, C. E., 74n, 75
σ, *see* standard deviation
σ-age, 69
significance, 81, 90–101
   see sample, small and large
simplicity, 5, 11, 55n, 60, 85–6, 89
Slater, J. C., 75
slide-rule, generalised, 105n
Smith, C. A. B., 71n
smoking, 41n
smoothness, 45, 85, 88, 89
sociology, 83
solipsism, 11
so-much-or-more method, 93–4, 96, 97
space, finite-dimensional, 9, 90, 96–7, 99
   see volume
space of *functions*, 61, 85
"spread", 55
standard deviation, 54
   see variance
standard measure, 56
star magnitudes, 64
state of mind, 2
statistic. A numerical function of observations. Thus the word "statistics" has two meanings

117

statistical—
  hypothesis, *see* hypothesis, statistical
  mechanics, 8, 75
  theory of probability, 6, 10
statistics—
  and probability, 59–61, 76–103
  definition, 76
  descriptive, 76
  of statistics, 86
  predictive, 76
Stieltjes, T. J., 51
Stirling's formula, 57
subjective, *see* objective
subjectivity, *see* objectivity
substantially right, 46
" success ", 6, 7, 29
suggestions, 34–6, 45, 60
  see axioms, rules and suggestions
sum of random variables, 51–2, 58, 59
  see convolution
super-population, 85n
superstition, 83
support, 63n
symmetric function, elementary, 38n
symmetry, 8, 17, 37, 41, 90, 96

T

T1, T2, etc., *see* theorems
tables, 102
  see contingency tables
" tails ", 60n, 87–8
tautology, *Preface*
  see probability, tautological
Tchebycheff's inequality, 57
telepathy, *see* extra-sensory perception
tests, *see* trials; sequential tests; significance
theorem, central limit, 57
theorems—
  T1 to T20, 22–8
  T21, T21A, 52
  T22, T23, 63
  T24, 79
  see mathematical theorems; Bayes' theorem; probability, fundamental theorem of; factors, weighted average; Borel's theorem
theories of probability, classification, 6–12
  see theory of probability
theories, scientific, *see* scientific theories; *see under* hypotheses
theory, abstract, 5, 19–30
  We use the word " theory " in several senses
theory of probability, 1, 3, 31, 34, 76n
  classification of our, 11–12
  purposes of, 3, 48–9
  see frequency theory of probability; statistical theory of probability
time, variations of beliefs with, 3
time-saving, 77, 95
Tintner, G., 48
Todhunter, I., 54
tolerance limits, 102
transitive body of beliefs, 14n
transmission lines, 64
trials, 6, 28, 73, 78
  see experiment, etc.
  expected number of, 65, 73
true value of a physical magnitude, 89
truth tables, 28
Tukey, J., 75
Turing, A. M., 63, 72, 73
  see computable numbers
types, theory of, 41n, 89n
typical value, 54
typicalness, fallacy of, 67

U

unbiased estimate, 103
universe, *see under* imaginary
unobservables, 30, 36, 48
unreasonable, 5, 14, 32, 49
  see reasonable
urgency *versus* importance, 95
Uspensky, J. V., 6n, 29n, 73
utility, 53–4
  and Ramsey, 10

INDEX

utility, and sequential tests, 65
    judgment of, 48
    neglect of, 102
    of alternative probability techniques, 89n
    of approximate methods, 77
    of gambling, 54
    of scientific theories, 10, 40

V

vagueness, 66, 97, 98n
values, scale of, *see* utility
variable, random, 50
variance, 54, 60
Venn, J. A., 6, 62n
verification of the theory, 39–40
    see justification
volition, 41n
volume, 55, 61, 90, 95–7, 104
    see function space

W

Wald, A., 7n, 64–6
Watson, G. N., 104n
wave function, 42
wearing out, 80
weather forecasts, 49
weighing evidence, 62–75
weight of evidence, 48, 63
    and chi-squared, 91–2
    expected, 72, 73
    relative, 71, 75
Weyl, H., 58n
wheel, rotation of a, 57
Whittaker, E. T., 104n
Wiener, N., 75
Wilks, S. S., 57, 76n, 87n, 101n, 102
Wintner, A., 88
Wright, G. H. von, 21n

Y

" You ", *Preface*, 2

Date Due

DEC 4 1980			
JAN 15 1987			
FEB 10 1988			
FEB 10 1998			

QA273 .G65
Good, Isidore Jacob
  Probability and the weighing of
evidence